ON LOCATION

OXFORD ARISTOTLE STUDIES

General Editors

Julia Annas and Lindsay Judson

Other titles in the series

Aristotle on Meaning and Essence
David Charles

Aristotle's Theory of Substance
The Categories *and* Metaphysics *Zeta*
Michael V. Wedin

Order in Multiplicity
Homonymy in the Philosophy of Aristotle
Christopher Shields

Aristotle's *De Interpretatione*
Contradiction and Dialectic
C. W. A. Whitaker

On Location

Aristotle's Concept of Place

BENJAMIN MORISON

CLARENDON PRESS · OXFORD
2002

OXFORD
UNIVERSITY PRESS

Great Clarendon Street, Oxford OX2 6DP

Oxford University Press is a department of the University of Oxford.
It furthers the University's objective of excellence in research, scholarship,
and education by publishing worldwide in

Oxford New York

Auckland Bangkok Buenos Aires Cape Town Chennai
Dar es Salaam Delhi Hong Kong Istanbul Karachi Kolkata
Kuala Lumpur Madrid Melbourne Mexico City Mumbai Nairobi
São Paulo Shanghai Singapore Taipei Tokyo Toronto

with an associated company in Berlin

Oxford is a registered trade mark of Oxford University Press
in the UK and in certain other countries

Published in the United States
by Oxford University Press Inc., New York

First published 2002

British Library Cataloguing in Publication Data
Data available

Library of Congress Cataloging in Publication Data
Morison, Ben (Benjamin)
On location: Aristotle's concept of place / Ben Morison.
p. cm. — (Oxford Aristotle studies)
Includes bibliographical references (p.) and index.
1. Aristotle. Physics 2. Place (Philosophy) 3. Physics—Philosophy. I. Title II. Series.
Q151.M67 2002 2001054557
114—dc21

ISBN 0-19-924791-9

1 3 5 7 9 10 8 6 4 2

Typeset by John Waś, Oxford
Printed in Great Britain
on acid-free paper by
Biddles Ltd., Guildford and King's Lynn

Acknowledgements

THIS book is a revised version of my doctoral thesis. It simply would not exist were it not for my supervisor Michael Frede. His guiding overview and gentle encouragement while overseeing the work, and afterwards, have been inspirational. I have a longer-standing and ever increasing debt to Jonathan Barnes. He taught me how to do philosophy, and in particular showed me how to read Aristotle. Anyone familiar with ancient philosophy will know how fortunate I am to have had these two great men as teachers.

It was while staying with Jonathan and Jennifer Barnes at Les Charmilles that I first resolved to write about places; I want to take this opportunity to thank them for their hospitality then and on many other occasions.

My two doctoral examiners, Edward Hussey and David Sedley, were kind enough to provide me with many pages of suggestions for improvement, all of which were wide-reaching. I hope I have done their comments justice, as well as those made by an anonymous reader at Oxford University Press. Latterly, Hendrik Lorenz has greatly helped me clarify and improve my ideas. Others to whom I have incurred debts during the writing of this book include Corine Besson, George Boys-Stones, Kevin Mulligan, Hanna Pickard, Ian Rumfitt, Richard Sorabji, Christopher Taylor, the participants in my seminar on *Physics* IV at the University of Geneva, and the audiences at Oxford, Durham, and London who heard and picked apart the *Reader's Digest* version.

The bulk of the revision of the thesis was done while I was a Postdoctoral Fellow of the British Academy. I held the fellowship at Corpus Christi College, Oxford, as a non-stipendiary Junior Research Fellow. I should like to thank the President and Fellows of both institutions for bringing about the perfect conditions for work, and my Academy mentor, Susanne Bobzien, for her help, encouragement, and beautiful trills. The final touches to the revision were done at The Queen's College, Oxford, and I should like to thank the Provost and Fellows for their generous welcome.

Finally, I must thank my family, and Frederika, for their constant support during rather difficult times. They know how much I owe them.

August 2001 BCAM

Contents

Introduction

NOT very many people have defended Aristotle's account of place. Virtually no ancient or medieval philosopher accepted it, although nearly all of them felt the need to refute it—implicitly or explicitly—when giving their own theory.[1] Even the loyal Theophrastus, Aristotle's pupil and successor as head of the Peripatos, perceived an easy target and trained his guns on it, or so the story goes.[2] And if scholars have been quick to believe that story, it is precisely because they have thought that Aristotle's account is inadequate. In this book I aim to set out Aristotle's account of place in *Physics* IV 1–5 as clearly as possible, hoping to rehabilitate it as a piece of philosophy.

As a piece of philosophy, however, it belongs to Aristotle's *Physics*, and hence must be understood in the context of the general aim of that book. The key essays on cause (or explanation), chance, teleology, change, infinity, place, void, and time, which form the core of *Physics* II–IV, are supposed to articulate rather precisely what these basic concepts are, since they are those which a scientist might use in systematizing the knowledge he gains about the natural world. In this sense, 'the *Physics* ranks itself not with physics, in our sense of the word, but with philosophy'.[3] These elucidations are conducted in part using empirical data, in part from a more philosophical standpoint. For instance, Aristotle takes certain facts about the natural motions of the elements (earth goes down, fire goes up) as *conditions* on his account of place, such that

[1] See Sorabji [26], 199–200, for the general influence of Aristotle's account on other philosophers; for its influence on Epicurus see Inwood [60], with Sedley [61]. For the Middle Ages see Grant [68] and Lang [70].

[2] See Sorabji [59] and [26], chs. 11–12. Algra [56] and [21] contribute further to the debate. My own view is that Theophrastus did not *criticize* Aristotle's account—cf. Sedley [65], 140 n. 1, and Sharples [15]. For more on the reception of Aristotle's theory of place by Peripatetic philosophers, see the bibliography, under 'Peripatetic places'.

[3] Owen [75], 88. Part of Owen's argument is that Aristotle works with puzzles stemming from Plato's *Timaeus* and *Parmenides*, and Zeno of Elea—these sources (especially the latter two) are hardly empirical scientific studies. Wieland [79] defends the view that the *Physics* is a work of conceptual analysis.

his definition of what a place is must allow for there to be a distinction between above and below,[4] and his prima facie arguments for the existence of places come from empirical observation of the natural motion of the elements and of the phenomenon of replacement.[5] On the other hand, he makes trouble for the existence of places with the purely philosophical puzzle of Zeno's paradox of place.[6]

Why does Aristotle bother to elucidate and articulate the concept of *place*? I argue in Chapter 1 that it is because of the importance of places to motion, i.e. *change* of place. For various reasons, Aristotle takes motion to be the central or paradigm kind of change, and change is central to nature. The natural scientist cannot do without the concept of change; a huge number of scientific theorems make use of verbs of motion and change. Consider, for instance, the importance of verbs of motion in biology—one very important principle of differentiation among animals is how they *move*: do they walk, crawl, swim, or fly? If biology is a respectable discipline—if, that is, knowledge of its axioms and theorems will confer understanding of what living things are and how they operate—then the natural scientist (the *phusikos*) will have to be sure that the concept of local motion can be satisfactorily defined and does not involve any surprise paradoxes or collapse under close scrutiny.

A modern parallel might help to bring home the significance of this. Before properly laying out the science of kinematics, we need to know some basic logical facts about motion: whether there is such a thing as absolute motion; if not, relative to what motion should be measured, etc. Frege thought that disputes about relative and absolute motion 'can of course be settled quite easily by recognising that an expression of the form "*a* is in motion" is incomplete, and by substituting for it an expression of the form "*a* is in motion relative to *b*"'.[7] The 'theoretical unsuitability [of the expression "*a* is in motion"] is all the more readily overlooked as it helps one to sidestep many a difficulty without being noticed'.[8] A proper understanding of locutions to do with motion

[4] *Phys.* IV 4, 211^{a}3–4. Cf. *Phys.* III 5, 205^{b}31–5, where Aristotle argues that the existence of an infinite body would rule out such a distinction, and hence that there is no infinite body.

[5] See Chapter 1, 'The existence of places (1): replacement', pp. 20–5.

[6] See Chapter 3 below.

[7] Frege [103], 123 (original, 145–6).

[8] Ibid. 124 (original, 146).

will reveal what the proper concepts for the scientific theory of motion are.

Aristotle is not concerned with the question of relative and absolute motion,[9] but there is an equally formidable group of problems with which he has to do battle, namely Zeno's paradoxes of motion. If they cannot be defeated, then the very concept of motion is under threat, and the scientist would have to stop using verbs such as 'walk', 'fly', etc. For this reason, Aristotle devotes some time in the *Physics* to solving them.[10] But equally, since motion is change of place, if it turns out that it is impossible to say what a *place* is, or to provide an account of what it is to say *where* something is, then we should stop employing the concept of motion when doing science; those theorems using vocabulary of motion will be meaningless, and hence knowledge of what they say will be impossible, and will not confer any understanding of nature.

However, it is not only the involvement of the concept of place with that of motion which is important. In the course of reporting observations or constructing theories about the natural world, the natural scientist will state *where* natural bodies, their parts, and other such things are, and draw conclusions from these statements. Moreover, there are at least two principles which the natural scientist will use which involve the concept of place even more directly: 'No two bodies can be in the same place at the same time',[11] and 'no body can be in two places at once'. It is clear that the concept of place must be able to be elucidated and articulated satisfactorily if the natural sciences are to be justified in making use of these principles.

Moreover, we cannot take it for granted that the concept of place will turn out to be acceptable just because we habitually talk of where things are, and refer to places. The only definition of 'void' that Aristotle accepts, namely 'a place in which there is no body',[12] is such that nothing actually does, or perhaps could, satisfy that predicate—so no natural science should make appeal to the void, e.g. in an explanation of how things move. It would be a disaster if a similar fate were to befall places. It is in this spirit that we should understand the questions that Aristotle poses

[9] See Chapter 5, 'Relative and absolute place', pp. 169–71.

[10] *Phys.* VI 9; cf. also *Phys.* VI 2 and VIII 8.

[11] Cf. *Phys.* IV 1, 209^{a}6–7.

[12] Cf. *Phys.* IV 7, 213^{b}33.

concerning place: *whether* it exists or not, *how* it exists, and *what* it is.[13]

Aristotle does not think that places are among the primary furniture of the world. He thinks that they are in some sense parasitic entities, an 'epiphenomenon' of the way the world is. This sort of view is classically Aristotelian, familiar from the *Categories* and *Metaphysics*. There are substances, which are the primary things, and then there are the properties which substances have, like qualities (colours etc.), quantities (size, shape, etc.) and so on, including *places*. These properties of substances are the sort of thing you mention when you give canonical answers to certain questions *about* substances. So, for instance, the question 'What is Socrates like?' might be answered 'He is pale'. This answer is a predication of quality—it says what Socrates is like—and the item introduced in this answer, namely pallor, is a quality.[14] However, you cannot have pallor without substances to *be* pale; for pallor to exist is for there to be substances which are pale. Substances do not depend on some other sort of item for their existence, but qualities so depend on substances.

If one is interested in places, and one is adopting this sort of framework, the natural questions to investigate are *where*-questions, and it is no surprise that Aristotle has a category of predication called *where*. The answers to where-questions will be predications which say *where* something is, so we might have the following simple thought: the items specified in these answers are *places*. (Then 'place' would be the abstract noun corresponding to 'where', as 'quantity' corresponds to 'how much'.) In the *Categories* Aristotle gives two examples of predications which say where something is, i.e. according to our approach, two 'canonical' answers to where-questions: 'in the Lyceum' and 'in the agora'.[15] Suppose the question had been 'where is Coriscus?'. The items we have introduced in answering this question (in saying *where* Coriscus is) are the Lyceum or the agora, so the Lyceum and the agora are *places*.[16] Here, though, things get complicated. Is

[13] *Phys.* IV 1, 208^a28–9: *εἰ ἔστιν ἢ μή, καὶ πῶς ἔστι, καὶ τί ἐστιν.*

[14] Connecting the theory of the categories with certain questions and their answers is natural, since most of the ten categories have *interrogative* particles as their names, *viz.* the categories of what, what is it like, how much, in relation to what, where, when. See Kahn [116] for further discussion.

[15] *Cat.* 4, 2^a1–2: *ἐν Λυκείῳ, ἐν ἀγορᾷ*. Cf. *Phys.* IV 11, 219^b21.

[16] There is an aggravating difficulty here, which is that in the *Categories* Aristotle

it really true that for the Lyceum to exist is for certain other things—substances—to exist? It seems not. But equally, for it to be a *place*, there must be something *of which* it is the place, or *of which* it could be the place. For places to exist is for them to be occupied—by other things. *Qua* place, the Lyceum depends on its occupants.

That simple thought, that the items specified in answering where-questions are places, is *too* simple. Aristotle believes that everything is somewhere, but that *not* all things have places. Wisdom, for instance, although *somewhere*, is not in a place; it is rather that the thing in which wisdom 'resides' (or whatever word is appropriate) is in a place. So wisdom may be somewhere, namely in Socrates, but it is Socrates who is in a place. The upshot is that places are not, *unqualifiedly*, answers to where-questions—only to *certain* where-questions.[17] The precise identification of which things have places is therefore one of Aristotle's prime concerns in *Physics* IV 1–5, as is the exploration of how exactly those things which are somewhere but which do not have places (e.g. properties) are somewhere in virtue of being related in some way to those things which *do* have places.[18] Only when we have a proper understanding of what it is for something to be somewhere can we understand what it is to answer where-questions properly, i.e. what it is to say where things are.

Predications which say where x is are suitable answers to questions such as 'where is x?' or 'where is x ϕ-ing?'.[19] Aristotle appears to think that canonical answers to such questions have the form 'x is in y',[20] for the only examples of a predication in the category of

speaks of places (and times) in connection with the category of *how much* (*Cat.* 6, 4^b24–5; cf. 5^a6–14, 6^a11–18), not the category of *where* (or *when*). However, this hardly shows that places and times are not connected with the categories of *where* and *when*, since at no point in the *Categories* does Aristotle actually give any details about the categories of *where* or *when*. Ackrill is surely near the truth when he points out that when discussing the category of *how much*, Aristotle 'lists and groups the *owners* of quantitative properties' ([83], 91; his italics). It is perfectly possible for places to be those things introduced in answering where-questions, and yet for them to have quantity. Cf. questions such as 'How much place/space does x take up?'

[17] For more on this see Chapter 1, 'Somewhere', pp. 15–20.

[18] See Chapters 2 and 3 for more on this.

[19] Presumably because asking where x was ϕ-ing is equivalent to asking *where* x was *when* x was ϕ-ing.

[20] Or rather, of course: 'x ἐστι ἐν y'. Higginbotham [110], 201, seems to suggest

where in the *Categories* are 'in the Lyceum' and 'in the agora', and in the *Physics* Aristotle talks throughout as if it were obvious that something's place is what it is *in*. On the face of it, this is rather strange. If I ask where something is, and someone replies that it is *to the right of* the toaster, there seems no obvious reason to deny that I have been told where it is. Other than '*x* is in *y*', two-place predicates which one might imagine were suitable for saying where something is include '*x* is to the right of *y*', '*x* is to the left of *y*', '*x* is in front of *y*', '*x* is behind *y*', '*x* is above *y*', '*x* is below *y*', '*x* is north of *y*', '*x* is east of *y*', '*x* is south of *y*', '*x* is west of *y*', '*x* is next [in line] to *y*'.[21] All these can be *quantified* as well, in this sense: one can construct predicates such as '*x* is twenty miles north of *y*' etc., and '*x* is second in line from *y*'[22] etc. Moreover, why exclude three-place predicates such as '*x* is between *y* and *z*'? These all seem perfectly good candidates for featuring in predications which say where *x* is.

Aristotle nowhere defends his restriction to 'in', but it seems to me that there are two main reasons for it.

The first is that in Greek, when you say that some physical body or other is somewhere, as opposed to nowhere, you can say idiomatically that it is 'in (a) place' ('*en topōi*'). This surely led Aristotle to reflect that something's place must be something *in* which it is, and hence that something's place is, roughly speaking, its *surroundings*. Thus he examines the things in which something *x* is, to see if any qualifies to be *x*'s place, and if so, how many, and whether they all have *equal* title to be *x*'s place.

The second more or less follows on from the first. Suppose someone asks where something *x* is. Then the reply (if it is of the form '*x* is in *y*') may be in the right spirit, but somehow not precise enough. Someone may answer '*x* is in the house', but this may (depending on what *x* is) provoke the rejoinder 'where in the house?—be more precise'. A further, more precise, answer might be '*x* is in the kitchen'—that second answer is more precise because anything in

that in English, answers should be of the form '[*x* is] at *y*': '*where* incorporates *at*, and is effectively equivalent to *at which place*'. 'At' is certainly one of the meanings of the Greek ἐν.

[21] Galen gives an example of a proposition which concerns *where* something is which does not involve 'in': ἔνιαι [sc. τῶν προτάσεων] δὲ ὑπὲρ τοῦ ποῦ "δεύτερος ἀπὸ γῆς ἐστιν ὁ ἥλιος"—'the sun is second in line from the earth' (*Institutio logica*, II 1). However, I know of no other similar examples in Greek philosophy.

[22] Cf. the passage from Galen quoted in the previous note.

the kitchen is in the house, but not vice versa. It is, however, still imprecise in this sense: other things might also be in that room; the respondent has not yet narrowed the possibilities down as much as possible. If there were an item *p* which contains nothing more than *x* itself, then specifying *p* would be the most precise answer to the question 'where is *x*?'

This 'nesting' of ever more precise places is utterly characteristic of Aristotle's discussion in *Physics* IV 1–5,[23] and I think he was struck by the fact that it is not possible to narrow down where something is by being ever more precise about what it is to the right of (for instance), not even if you say what it is *immediately* to the right of. For a start, there are certain difficulties, to which Aristotle refers,[24] in how to understand '*x* is to the right of *y*' and similar relations. But even if we overcame these, '*x* is to the right of *y*' and similar relations take care of only one edge of something—they take in only one part of *x*'s surroundings. Here is an example. At the opera, Mr Smith is immediately to the right of me. But so is the Smith family. Yet Mr Smith and the Smith family occupy different places—Mr Smith has seat A 14, the Smith family seats A 14, 15, 16, and 17. However exact you are about what it is they are to the right of, you cannot distinguish *in this way alone* between where the Smith family is and where Mr Smith is. Here is another example, using the relation '*x* is above *y*'. A small building is demolished and a tower block built on the same site. The building *was* immediately above the same thing as the tower block *is* now (a particular patch of ground). But the tower block is not exactly where the building was;[25] for instance, it is nearer the sun than the building was (where you are makes a difference to how near you are to the sun—this is something the scientific study of the natural world should recognize). Saying what *x* is above is not to tell the whole story about where *x* is—it is not to specify fully *x*'s surroundings.[26] Again, suppose that the universe were infinite,

[23] Cf. *Phys.* IV 2, 209^{a}31–^{b}2; 4, 211^{a}23–9; 5, 212^{b}20–2.

[24] *Phys.* IV 1, 208^{b}14–22. See Chapter 1, 'The six dimensions', pp. 35–47.

[25] Aristotle does not think that two things which differ in size could ever occupy (at different times, of course) the same place. When something grows, it changes place, according to him.

[26] Perhaps Aristotle thinks that '*x* is to the right of *y*' is *in some sense* a predication which says where *x* is; if you reply '*x* is to the right of *y*' when someone asks 'where is *x*?', you have *in some sense* answered the question. The natural suggestion that this is a predication in the category of *in relation to what* has, unfortunately, no basis in

and that there are no infinitely large bodies. If you know where y is, then knowing that x is in y is (infinitely!) more helpful than knowing that x is to the right of y, or above y. The fact that x is in y immediately narrows down the possibilities of where x is, so to speak. Even if the universe is not infinite, we can still appreciate this *formal* property of the relation. The relation 'x is in y' is therefore suitable for narrowing down where x is; this is presumably one reason why Aristotle uses it to the exclusion of other relations.[27]

It is not possible for one body to be in two different places at the same time (this depressing thought has surely struck many of us often). In this respect, bodies are different from items such as wisdom, which can simultaneously be somewhere (in Socrates) and also somewhere else (in Plato); indeed, this may be one of the important ontological or metaphysical differences between bodies and properties. Conversely, distinct bodies cannot be in the same place at the same time, whereas distinct properties can be realized in the same item. However, these two principles, that one body cannot be in two places at the same time, and distinct bodies cannot be in the same place at the same time, seem to be vitiated by our perfectly ordinary way of talking. If x is in the house, and more precisely, in the kitchen, and more precisely, in that thing p which contains nothing more than x, then the house, the kitchen and p *all* seem to be places of x. In that case, x is in many different places (p, the kitchen, the house) at the same time. Equally, if there were to be something else, y, in the kitchen at the same time as x, then x and y would be in the same place (the kitchen, the house) at the same time.

Aristotle wavers between the following two solutions to these difficulties. (i) Distinguish between the 'primary' or 'proper'[28] place of x, i.e. p, and x's other places, and then recast the two principles

the discussion of relatives in the *Categories*. Presumably Aristotle thinks that these other predications sometimes constitute adequate answers to where-questions because they do tell you *something* about the surroundings of the item in question; i.e. that they do help you build up a picture of those surroundings and hence identify where the thing in question is.

[27] For another reason see Chapter 2, 'Conclusion: being in a place', pp. 78–80.

[28] πρῶτος (*Phys.* IV 3, 210^{b}24, and 4, 211^{a}2), and ἴδιος (2, 209^{a}33). Sextus Empiricus will later refer to the 'strict' (κυρίως, *PH* III 119), 'circumscribed' (κατὰ περιγραφήν, *M* X 15), or 'accurate' (κατ' ἀκρίβειαν, *M* X 95) sense in which something is x's place. See Burnyeat [43] for Sextus on places.

as 'two bodies cannot be in the same *proper* place at the same time' and 'it is not possible for a body to be in two different *proper* places at the same time'. (ii) Regard the predications 'x is in the kitchen' and 'x is in the house' as imprecise specifications of the (unique, proper) place of x; then the fact that x is in p and x is in the kitchen does not imply that x is in two different places at once, nor does the fact that x and y are both in the kitchen imply that x and y are in the same place simultaneously. Aristotle seems to opt for (i), although it is sometimes easier to interpret his account of place in terms of (ii).[29] The crucial point, however, is that on either account, *proper places* are the key to an understanding of the principles 'two distinct bodies cannot be in the same place at the same time' and 'one body cannot be in two distinct places at the same time'. It is a perfectly familiar feature of our everyday practice of saying where things are that there are more or less precise specifications of where things are, and proper places are simply the items introduced in the most precise specification possible of where something is; recognition of them is therefore—in some sense—inherent in that practice. For this reason, much of *Physics* IV 1–5 is an attempt (i) to show that there is such a thing as a most precise specification of where something is, and therefore such a thing as the proper place of something, and (ii) to say what the proper place of something actually is.

I have tried in the foregoing to indicate broadly how I think Aristotle's discussion is best understood. One thing is worth emphasizing. I do not think that Aristotle tries to give a *semantics* for where-questions, or anything of that kind. There are a bewildering variety of different where-questions, and different answers are given in reply to them depending on all sorts of parameters, such as the beliefs that the respondent has about how much the questioner knows, and so on; there may even sometimes be reasons for producing incorrect, partial, or misleading answers. Rather, Aristotle aims to bring out and elucidate the presupposition *behind* such questions, namely that things are *somewhere*, and to analyse that crucial part of our practice of locating things which consists of stating what they are *in*. He shows that it is a practice which is free from paradox, and which has at its core the notion of a proper place, of which he provides a definition.

[29] See Chapter 3, 'The problem (*b*): What is unacceptable about the series?', pp. 92–6.

We might wonder where the keys are, where the light-switch is, where the sun will be at noon tomorrow, where this smell or sound is coming from, why the petals of plants are where they are, where the liver is situated, whether flying fish fly or glide, and so on. *Physics* IV 1–5 tells us exactly what it is for bodies, causes, parts, etc. to be somewhere, and shows us that there is no paradox or absurdity which arises in the concept of motion in so far as it is understood as change of *place* (any problems there may be will come from its being *change* of place). The practice of locating things—common to both the natural scientists and us—is elucidated, articulated, and vindicated.

1

Places, Natural Places, and the Power of Place

The importance of motion

ARISTOTLE opens his discussion of places with a set of three programmatic questions which need to be answered by the student of nature: (i) whether places exist or not, (ii) how they exist, and (iii) what they are. The first question is clearly the most pressing, because the second and third questions presuppose that the answer to it is that places *do* exist. But why should the first question even arise, or interest the student of nature? Why should the student of nature give a second's thought to *places*?

Places are important to the study of nature, Aristotle tells us, because (*a*) 'everyone supposes that those things which are are somewhere',[1] and (*b*) 'the most common sort of change, and that which is most properly so called, is change in respect of place, which is called local motion'.[2] The second of these two reasons serves to put the discussion of places in the wider context of Aristotelian physics. For at the start of *Physics* II 1 Aristotle states that natural bodies are those which have in them 'a principle of change and rest'.[3] I have translated the Greek word *kinēsis* here by 'change', although it is often rendered 'motion'. However, it does not have the same restricted meaning as that English word. In English, something which is changing colour is not said to be *moving* (although perhaps one can say that the traffic light has *moved* from being red to being green); nor is something which is growing said to be *moving*. But *kinēsis* includes more than just motion, for Aristotle goes on to

[1] *Phys.* IV 1, 208^{a}29: τά τε γὰρ ὄντα πάντες ὑπολαμβάνουσιν εἶναί που.

[2] *Phys.* IV 1, 208^{a}31–2: τῆς κινήσεως ἡ κοινὴ μάλιστα καὶ κυριωτάτη κατὰ τόπον ἐστίν, ἣν καλοῦμεν φοράν.

[3] *Phys.* II 1, 192^{b}14: ἀρχὴν ἔχει κινήσεως καὶ στάσεως.

specify that the 'motions' in question are those in respect of place, growth and diminution, and alteration.[4] At *Physics* V 1–2 Aristotle restricts the range of *kinēsis* to just these changes, preferring to use *metabolē* for change even more generally, including generation and destruction (or coming-to-be and passing-away, as they are sometimes known).[5] Elsewhere in *Physics* I–IV he apparently includes generation and destruction in the range of *kinēsis*, equating it with *metabolē*.[6] However, when Aristotle says that natural bodies are those things which have a principle or cause of change and rest, he specifies which changes he means. Natural bodies are those bodies which are able to initiate and stop these sorts of change because of *what they are*. So trees change and stop changing in certain ways—they grow and stop growing, they blossom and stop blossoming, etc.—and do so because they are trees. Beds, on the other hand, do not do any of those things; and if a bed does undergo a change which originates in it (for instance, if it rots), then it has done so not in virtue of being a bed, but in virtue of being made of wood—not, therefore, in virtue of what it *is*, but in virtue of *that out of which it is made*.

Natural bodies are those which contain a principle of *kinēsis* and *stasis*, and the changes in question are change of place, growth and diminution, alteration, and substantial change. None the less, it is not entirely misguided to translate *kinēsis* by 'motion', for its *primary* meaning is indeed 'motion', i.e. change of place. So, at *Theaetetus* 181 C–D, Plato has Socrates lead Theodorus from agreeing that change of place and moving around in the same place are examples of *kinēsis*, to agreeing that *other* changes (of colour, hardness, etc.) are *also* kinds of *kinēsis*. Now, Aristotle reiterates at the start of book III of the *Physics* that nature is a principle of *kinēsis* and *metabolē*.[7] He goes on to say that place, void, and time are necessary for *kinēsis*[8]—a nod towards book IV of the *Physics*, which treats

[4] *Phys.* II 1, 192^b14–15: *τὰ μὲν κατὰ τόπον, τὰ δὲ κατ' αὔξησιν καὶ φθίσιν, τὰ δὲ κατ' ἀλλοίωσιν.*

[5] See e.g. *Phys.* V 1, 225^b7–9.

[6] e.g. *Phys.* III 1, 210^a8–9: *ὥστε κινήσεως καὶ μεταβολῆς ἔστιν εἴδη τοσαῦτα ὅσα τοῦ ὄντος.* Aristotle *explicitly* includes generation and destruction as types of *κίνησις* at *Cat.* 14, 15^a13–14: 'There are six types of change: generation, destruction, increase, diminution, alteration, change of place' (*κινήσεως δέ ἐστιν εἴδη ἕξ· γένεσις, φθορά, αὔξησις, μείωσις, ἀλλοίωσις, κατὰ τόπον μεταβολή*). The restriction of the meaning of *κίνησις* in book V and its significance to the question of the division of the *Physics* (see n. 11 below) are discussed by Brunschwig [81], 32 ff.

[7] III 1, 200^b12–13: *ἀρχὴ κινήσεως καὶ μεταβολῆς.*

[8] III 1, 200^b21: *ἄνευ τόπου καὶ κενοῦ καὶ χρόνου κίνησιν ἀδύνατον εἶναι.* Aristotle

just these three topics, in that order. Place, void, and time may well seem to be necessary conditions of local motion, but they hardly seem to be necessary conditions of *every* change (although Melissus did think that no change at all would be possible without void,[9] perhaps because he thought that no change would be possible without locomotion, and that locomotion is impossible without void). None the less, the initial plausibility of the necessity of these factors for change in general comes from the fact that *kinēsis* applies first of all to changes of place, and only secondarily to *all* changes. Aristotle supposes that its *wider* meaning depends on its *primary* meaning in some way—it is not a matter of chance that *kinēsis* applies to changes of place and changes quite generally.[10] Thus, you cannot get totally clear about *kinēsis* in its wider senses without first getting clear about its primary sense, 'change of place'. (Similar remarks apply to *stasis*, whose primary meaning is 'standing still', i.e. 'not changing place'.) The study of change quite generally must give pride of place to a study of change of place, and so must contain a study of what places are. Hence, chapters 1 to 3 of book III are devoted to an account of *kinēsis* quite generally, and book IV opens with the remark that the student of nature should study places.[11]

It is to this that Aristotle alludes when he says that the student of nature should study places because 'the most common sort of change, and that which is most properly so called'[12] is change of place (for which he uses a single—precise—word, *phora*). Of all the kinds of change, locomotion is the one to which *kinēsis* most

must mean that these *seem* to be necessary—he cannot think that the void is *actually* necessary for κίνησις, for the simple reason that he thinks the void does not exist. Ross suggests that 'The sentence is under the governance of φασί "understood from" δοκεῖ (b16)' (Ross [3], 534); *previous natural philosophers* say that change is impossible without place, void, and time.

[9] See *Phys.* IV 6, 213^{b}12–14 (=30 A 8 DK).

[10] This is typically Aristotelian, and is connected with the doctrine of the πολλαχῶς λεγόμενα. See below, Chapter 2, 'One thing is said to be in another in many ways'.

[11] It is worth underlining that the division into books of what we call the *Physics* is non-Aristotelian, and, furthermore, that the *Physics* consists (in some sense) of at least two works, perhaps 'On First Principles' (Περὶ ἀρχῶν) and 'On Motion' (Περὶ κίνησεως). But scholarly debate has raged over whether the crucial division should come between books IV and V (in which case 'On Motion' consisted of books V, VI, and VIII, book VII being misplaced), or between V and VI (in which case 'On Motion' consisted of VI, VII, and VIII)—no one doubts that book IV belongs with book III. For more on the division of the *Physics* see Brunschwig [81], 28–31, and Barnes [80], 34–6, 59–61, and appendix.

[12] IV 1, 208^{a}31–2: κοινὴ μάλιστα καὶ κυριωτάτη.

properly applies. Aristotle wishes to explain how it is that the wider meaning of *kinēsis* depends on that more restricted meaning; his explanation is contained in the remark that change of place is the most common change.

This claim presumably amounts to the following: all changes involve change of place, but changes of place need not involve any other type of change. The claim is theoretical, and so needs to be backed up by a theoretical argument. That argument can be found at *Physics* VIII 7, 260a26–261a26, where Aristotle attempts to show that locomotion is the primary kind of change[13] (the candidates are change in respect of size, quality, and place). Aristotle produces a whole set of arguments, seemingly rather tangled up, aimed at establishing that locomotion is primary in each of the senses of 'primary' which he discerns.[14] The second and third of these are not important here; Aristotle shows that motion is temporally prior to all other changes (that is, before any change, something must move),[15] and primary in another sense, because according to him, there must be some one change going on continuously, but motion (in fact, circular motion) is the only change that could be going on continuously.[16] However, the argument for the primariness of motion in the first sense (apparently corresponding to the sense in which something is primary in being),[17] involves showing that any instance of any change somehow *involves* locomotion, but that there are instances of locomotion without necessarily one of the others.

Aristotle says (260b7–15) that a change in quality of something is actually just a condensation or rarefaction of its components, and condensation and rarefaction in a body are just cases of—presumably—parts of that body changing place. (Condensation and rarefaction also underlie generations and destructions, Aristotle adds.) Furthermore, anything which grows or shrinks actually changes place. Thus, Aristotle suggests that change of place is somehow involved in any kind of change; it has a *different* relation to each of growth–diminution, generation–destruction, and alteration, but is none the less present in all, hence common to all. It is perhaps overstating the case, but in the right spirit, to say

[13] Aristotle states in *Meta.* Λ 7 that motion is the primary form of change (1072b8, 1073a12), without argument.

[14] He lists them, confusingly, half way through the argument, at 260b17–19.

[15] 260b29–261a12.

[16] 260b15–29.

[17] 260b19: κατ' οὐσίαν.

that any change is a change of place of something, *described in a certain way*. (Any change undergone by *x* is such that either *x* or *x*'s parts change place.) What Aristotle says in *Physics* VIII can be used to show that any change whatsoever involves things changing place, and thus, since locomotion can occur without any of the other changes occurring, change of place is the most widespread change.

The fact that motion is the most common change explains how it is that *kinēsis* applies most properly to change of place,[18] for the argument that change of place is the primary change also shows that change of place is the most common change.[19] Aristotle's second reason for why the student of nature should study places is thus that change of place is central to *kinēsis*, i.e. change in general. In this way, book IV, which opens with the discussion of places, follows on smoothly from books II and III.

Somewhere

The first reason why the student of nature should study places is that 'everyone supposes that those things which are are somewhere',[20] in other words, that *something exists iff it is somewhere*.[21] This is put forward as a common-sense view (*everyone* supposes it), and amounts to something like this: something exists if and only if there is somewhere where it is, so if it is an object then it is somewhere in the world, if it is a property there is an object in which it resides, etc. This view might gain some plausibility from the fact that in Greek, as in English, there seems to be a locative use of 'is'. Roughly speaking,[22] it makes sense to qualify 'is' with locative adverbial phrases ('he is under the table', 'he is with the horses'), but not, for instance, with adverbs of manner ('he is slowly') or

[18] Hussey [5], 99, has an alternative suggestion for what Aristotle is getting at: motion is 'the "most general" (*koinê*) [kind of change] perhaps because even the otherwise changeless celestial bodies are subject to it'.

[19] Interestingly, Simplicius reports two variant readings: 'in some manuscripts *καὶ πρώτη* is written with *κοινή*, in others *πρώτη* instead of *κοινή*, as Eudemus writes' (*In Phys.* 522. 24–6 = fr. 72 Wehrli). Whether or not Eudemus' reading is correct, this confirms the close connection between change of place's being the most common change and its being the primary form of change.

[20] IV 1, 208^a29.

[21] 'Iff' is shorthand for 'if and only if'.

[22] See Kahn [115], 156–67, for all the necessary details and qualifications. What I say derives primarily from an observation he makes at p. 158.

instrumental adverbial phrases ('he is with the knife').[23] But this locative 'is' expresses *existence* in the locution '*there* is . . .'.[24] The idea is that in Greek, the existential use of the verb 'to be' either is identical with, or derives from, that locative use; it has the force of 'there is'.[25] When we say that something exists, we are saying that it is *somewhere*.

The interrogative particle 'where?' and the locative adverb of quantification[26] 'somewhere' are closely related. (This is particularly obvious in Greek, since the words for each are identical, bar accentuation.) Broadly speaking, it is possible to specify where x is in reply to the question 'where is x?' iff x is somewhere.[27] The item specified in answer to this question is, perhaps, x's place, or at least one of x's places. But if everyone supposes that something exists iff it is somewhere, then this seems to amount to thinking that something exists iff it is related to a place. This should be enough to whet anyone's appetite. Contrast, for instance, 'something exists iff it has qualities'. In Aristotle's view of things, this is not true—qualities themselves do not have qualities, yet they exist. If something exists iff it is related to a place, then places seem to be metaphysically on a par with substances, for Aristotle's official doctrine is that something *exists* iff it is somehow related to a substance.

There is none the less an important difference between substances and places: the things which are somewhere can *change* where they are, while retaining their identity. This is not the case with substances; if one substance becomes another then the first ceases to exist. A moment's reflection shows that it is precisely this metaphysical property of places which interests the student of nature. A rock falls naturally to earth, but it is one and the same rock

[23] We can qualify 'is' with temporal adverbial phrases, but only, it seems, when the subject of 'is' refers to something like an event ('The Battle of Waterloo was a long time ago'). 'That stone was yesterday' makes no sense.

[24] Cf. the French 'il y a'. How many other languages explicitly express existence with spatial adverbs?

[25] See Kahn [115], 164: 'a construction of the form '*N* is *PN*' with a literal local sense has an additional nuance which we describe in terms of existence or render into English by our locution "There is a such-and-such".' See also references in his index, s.v. 'Locative construction of copula, connected with existential sense, and with vital sense'.

[26] See Lewis [118] for discussion of this logico-grammatical category.

[27] This is not quite right, since presumably no language exists in which all places have a name, so we would have to supplement the expressive resources of the language, or perhaps use demonstratives. Aristotle is perfectly happy to use demonstrative phrases such as 'that place' (cf. *Phys.* IV 2, 209^{a}35: ἐν τῷδε τῷ τόπῳ).

which was up there, and is now down here. A study of nature could not get off the ground without recognizing that it is the same body, even when it actualizes its capacity to change (in this case, move). So—perhaps—something exists iff it is somewhere, but there is no single place particular to it which something must keep in order to retain its identity.

Up until now, I have assumed that being somewhere is being in a place. Elsewhere in IV 1 Aristotle remarks that most people suppose that everything is 'somewhere, i.e. in a place'.[28] But perhaps after all it is not true that the *only* way to be somewhere is to have a place, that the *only* way to answer the question 'where is *x*?' is by specifying a place. Perhaps it is right that the common-sense view is correct, and *properties* are somewhere, and that there are predications which say where they are, in what they reside. But surely this need not imply that properties thereby have a *place*. In fact, Aristotle nowhere defines 'somewhere' (*pou*), and appears to waver between

(1) *x* is somewhere iff *x* is in a place

and

(2) *x* is somewhere iff *x* is in something.

Pou in the first 'strict' sense appears at, for instance, IV 5, 212^{b}8–10 ('the universe . . . is not somewhere as a whole, i.e. not in some place, since no body surrounds it'), and in the second sense in the same chapter, at 212^{b}27 ('places exist and are somewhere, not as in a place, but in [another] way').

Eudemus and Alexander of Aphrodisias both point out[29]—rightly—that the second way of taking 'somewhere' must be read between the lines of Aristotle's treatment of Zeno's paradox of place,[30] and it is safe to suppose that Aristotle in fact would take the second of these definitions to be the correct one. (He thinks, however, that (1) is the *primary* sense of *pou*.) The things which are *in a place* are—with certain important exceptions to be introduced later—bodies. Properties etc. reside in these bodies; it is in virtue of being in

[28] IV 1, 208^{b}33: *πάντα εἶναί που καὶ ἐν τόπῳ.*

[29] Eudemus ap. Simplicius, *In Phys.* 563. 24–5 (=fr. 78. 22–3 Wehrli); Alexander, ibid. 563. 30: *ἰσοδυναμεῖ τὸ ἔν τινι τῷ που* ('"Being in something" has the same force as "being somewhere"').

[30] See below, Chapter 3, esp. p. 87 and pp. 97–102.

them that properties are somewhere. Thus, the common-sense view which Aristotle reports, that everything which exists is somewhere, prompts initially the thought that everything is in a place—that is the most obvious sense of 'somewhere'—and then the further thought that after all, being somewhere is more complicated than just being in a place; it is also to do with being *related* to something which has a place, by being *in* it. The importance of places even to this more generous sense of being somewhere is obvious.[31]

It is tempting to object that Aristotle says that people suppose *only* that the things which exist are somewhere, and that this leaves open, for instance, that the things which do not exist are also somewhere, that the things which exist are only a subset of the things which are somewhere. But I think that this would be over-pedantic. For Aristotle explains why people think that everything which exists is somewhere: 'For what is not is nowhere. For where is the goat-stag or the sphinx?'[32] If 'those things which are are somewhere' corresponds merely to

$$\forall x\,(\mathrm{E}x \rightarrow \mathrm{S}x),$$

then 'what is not is nowhere' should correspond to

$$\forall x\,(\neg\,\mathrm{E}x \rightarrow \neg\,\mathrm{S}x).$$

But, so understood, the first does not follow from the second,[33] and hence the second is not much use in justifying the first. Aristotle did not make a dull mistake; by 'everyone supposes that the things which exist are somewhere' he means that everyone thinks that all

[31] Even if we take 'somewhere' in the second, more generous, sense, it is not clear that 'those things which are are somewhere' comes out true. We have to face a rather special case: what is the unmoved mover in? One might point to *Phys.* VIII 10, 267^{b}9, where, according to the Oxford translation [1], Aristotle says 'the mover is at the circumference [of the universe]' (*ἐκεῖ ἄρα τὸ κινοῦν*—it is clear from the context that *ἐκεῖ* refers to the circumference of the universe). But perhaps things are not so clear (see Lang [69]). *Phys.* IV 5 (and, as Hussey [5], 99, points out, *De caelo* I 9, 279^{a}11–18) makes it quite clear that there is nothing outside the heavens, and so they are not *in a place*—neither, then, is the unmoved mover. This much at least is not a problem, for the unmoved mover could be *somewhere* in some other sense.

[32] *Phys.* IV 1, 208^{a}30–1. Elsewhere in *Phys.* IV Aristotle gives a slightly different explanation of how people have come to suppose this. 'They think that everything which exists is a *body*, and that every *body* is in a place' (IV 7, 213^{b}32–3: *τὸ ὂν σῶμα οἴονται εἶναι, πᾶν δὲ σῶμα ἐν τόπῳ*). Cf. Plato's Giants in the *Sophist*, who define body and being as the same thing (246 B 1: *ταὐτὸν σῶμα καὶ οὐσίαν ὁριζόμενοι*).

[33] Spotted by Simplicius (*In Phys.* 521. 15–20), and Philoponus (*In Phys.* 501. 25–502. 2).

and only the things which exist are somewhere. And they arrive at this view because they think that if something does not exist, you cannot say where it is (and if you cannot say where it is, it does not exist).

The view that everything which exists is somewhere is a common-sense view, metaphysically rather interesting, but also vague, because it is not clear what 'somewhere' really means. However, this view was also the subject of considerable philosophical debate. Zeno of Elea had already subtly drawn the proposition out into an infinite regression: if everything is somewhere, then surely even places have places, which have places, and so on. Aristotle clearly thought that this would constitute a *reductio* of some sort.[34] Moreover, consider the following passage from Plato's *Timaeus*:[35] 'looking at [space], we even dream[36] . . . and say that anything that is must needs be somewhere in some place and occupy some space, and that what is neither on earth nor somewhere in heaven is nothing'. The idea is that there *are* in fact some things which are not somewhere, and whose existence is extremely important, i.e. the Forms. We have lost concentration and not paid enough attention to the relation between the Forms and the world around us. We are tempted to think that everything which exists is somewhere, and that what is nowhere is nothing (does not exist),[37] but we are wrong; the Forms exist, but are nowhere.[38]

Plato appears to *gloss* 'somewhere' as 'in some place and occupying some space', in which case, as stated earlier, it is rather unclear whether everything that exists is somewhere. However, when Aristotle says that everyone supposes that everything is somewhere, he is, I take it, deliberately reminding us of Zeno's doubt and Plato's

[34] I discuss Zeno's paradox of place in detail in Chapter 3. Aristotle claims that the puzzle arises because of the vagueness of the meaning of 'somewhere'.

[35] 52 B 3–5: *πρὸς ὃ δὴ καὶ ὀνειροπολοῦμεν βλέποντες καί φαμεν ἀναγκαῖον εἶναι που τὸ ὂν ἅπαν ἔν τινι τόπῳ καὶ κατέχον χώραν τινά. τὸ δὲ μήτ' ἐν γῇ μήτε που κατ' οὐρανὸν οὐδὲν εἶναι.* I have more to say about this passage in Chapter 4, p. 117.

[36] Or: 'in relation to [space], we even dream with our eyes open'. See Taylor [17], 346, with Cornford [11], 192 n. 2.

[37] Note that Plato does not try to do more than restate that what exists is somewhere, when he says that what is nowhere does not exist. Aristotle, however, seems to want to explain why it is that people have thought that to be is to be somewhere, by starting from the thought that what is not is nowhere.

[38] As Aristotle points out at *Phys.* III 4, 203^{a}8–9. In *Phys.* IV 2, 209^{b}33–210^{a}2, he criticizes Plato for not realizing that in fact he is committed to giving the Forms a place. On this see Chapter 4, 'Plato's *Timaeus* 48 E–53 C'.

rejection.[39] The contention that everything which exists is somewhere is (1) an extremely interesting metaphysical thesis, (2) crucially vague because of the meaning of 'somewhere', (3) the subject of philosophical debate. To get clear about it and assess whether it is true or not, it is essential to get clear about *places*.

The existence of places (1): replacement

Of course, even if everyone says that everything is somewhere, they could be wrong: perhaps places just do not exist. And if places do not exist, then the subtleties in the meaning of *kinēsis* are just the products of our confused way of looking at the world. So Aristotle stops and asks what reasons there are for thinking that places actually do exist. He produces (208^{b}1–27) a series of prima facie arguments or considerations for their existence, and then a series of difficulties or puzzles which arise when we do posit their existence (209^{a}2–30). These puzzles constitute a challenge to any definition which is offered and Aristotle takes great pains to show how the puzzles are answered by his final definition.[40] However, the prima facie arguments for the existence of places reveal much about Aristotle's presuppositions. They start as follows:[41]

> That place exists seems clear from replacement: for where there is now water, there, when it goes out as from a vessel, air in turn is, and whenever some other body takes that same place. This, then, seems to be different from all the things that come to be in it, which move about—for water was formerly in that in which air is now—so that it is clear that place, and the space into which and out of which they moved in moving about, must be something other than either.

Aristotle's argument here proceeds via an analysis of replacement. In effect, Aristotle suggests that *x* has replaced *y* iff *x* is now where

[39] Thus when Aristotle claims that everybody supposes that those things which are are somewhere, the 'everybody' does not mean 'all philosophers' (rather as when one says 'everyone was at the party', meaning by 'everyone' only a rather select group of people). He means that this is a commonly held view.

[40] Although he does not entirely succeed. See p. 82 n. 9.

[41] 208^{b}1–8: *ὅτι μὲν οὖν ἔστιν ὁ τόπος, δοκεῖ δῆλον εἶναι ἐκ τῆς ἀντιμεταστάσεως· ὅπου γὰρ ἔστι νῦν ὕδωρ, ἐνταῦθα ἐξελθόντος ὥσπερ ἐξ ἀγγείου πάλιν ἀὴρ ἔνεστιν, ὁτὲ δὲ τὸν αὐτὸν τόπον τοῦτον ἄλλο τι τῶν σωμάτων κατέχει· τοῦτο δὴ τῶν ἐγγιγνομένων καὶ μεταβαλλόντων ἕτερον πάντων εἶναι δοκεῖ· ἐν ᾧ γὰρ ἀὴρ ἔστι νῦν, ὕδωρ ἐν τούτῳ πρότερον ἦν, ὥστε δῆλον ὡς ἦν ὁ τόπος τι καὶ ἡ χώρα ἕτερον ἀμφοῖν, εἰς ἣν καὶ ἐξ ἧς μετέβαλον.*

y was.[42] This is then expressed as *x*'s holding now the same place as *y* did. No part of the Greek word for replacement (*antimetastasis*) is etymologically linked to the Greek word for a place (*topos*). It derives from the verb *antimethistēmi*, whose middle means something like 'stand instead of', or 'stand in for'.[43] So Aristotle has seen that replacement is a question of one thing holding the same place that something else did before, but he has not done so by spotting that the word 'place' is present in the word 'replacement'.

When Aristotle says 'where there is now water, there . . . air in turn is', the word 'where'[44] seems to be a two-place sentential connective, as in 'I am going on holiday where my girlfriend rents a flat'.[45] Aristotle paraphrases 'where' by 'in the same place as', but there is no *logical* reason to assume that a connective of this sort involves quantification over some item such as a place. As Geach points out, the connective 'at the same time as' should not *necessarily* be viewed as embodying a quantification over time,[46] not least because it is often possible to use the word 'when', which on the face of it contains no such quantification and is etymologically independent of the word 'time'.[47] (Similarly, the use of the connective 'just as' does not justify an existential commitment to *ways*, even though it can be paraphrased 'in the same way as'.) Presumably, the same thing applies to the connective 'where'—it should not *necessarily* be thought to contain a quantification over places.

Aristotle was not committing a logical howler, however. In all probability, he was thinking of an argument of the following sort. If air is now where the water was, then where *is* the air? And where

[42] Perhaps one would wish to add the condition 'and nothing else was there in the meantime', *vel sim.*

[43] The verb is not uncommon, but the noun ἀντιμετάστασις appears here for the first time.

[44] 208^{b}2: ὅπου.

[45] See Geach [108], 315, where he notes that 'some medieval logicians' considered 'where' as an operator forming complex propositions ('local hypotheticals') out of simple ones. He claims (315–16) that a complex proposition containing 'where' cannot be joined to another sentence by 'where', and hence that its logical behaviour is so remote from normal connectives that it should not count as one. This interesting claim is not *obviously* right: consider, for instance, 'I nearly got run over where the road forks, where the hotel is where the coffee is terrible'.

[46] 'In particular, it is definitely wrong to analyse an unsophisticated simultaneity proposition, like "Peter was writing a letter and (at the same time) Jenny was practising the piano", in terms of what happened at some one time t—"For some time t, Peter was writing a letter at t and Jenny was practising the piano at t". Such a use of "at the same time" as we have here does not involve any reference to an apparatus or technique for telling the time (and still less, a reference to Absolute Time)' (Geach [108], 311).

[47] Ibid. 312.

was the water? In answering these questions, you need to mention something over and above the air and the water.[48] Aristotle's argument depends at this point on the supposition that when you want to say where the air is (and the water was), you should say *in what* the air is (and the water was).[49] Suppose then that there is something p such that at time t_1 water is in p and at time t_2 $(\neq t_1)$ air is in p. If p actually were that quantity of air or water we would have the absurd results either that the water was in itself at t_1 or that the air was in itself at t_2. Hence if x has replaced y, then saying where x is and saying where y was amount to the same thing, but the entity you mention when you say where x is, or where y was, can be neither x nor y. It follows that the place of x—the thing we mention when we say where x is—cannot be x itself.[50] Thus, places exist: they do not straightforwardly 'reduce to' their occupants. As Aristotle puts it: 'because of these considerations someone might suppose that place is something over and above bodies'.[51] There is perhaps another alternative—that the place of x is some aspect or other of x—but the crucial point is that places cannot be identified with their occupants.

At this point of the discussion, Aristotle has not fixed his terminology. He clearly thinks that something is *in* its place, but he also says that if x is where y was, then x 'holds the same place as' y did,[52] which suggests a different relation between an occupant and its place. Moreover, when he has shown that the item in question is different from the water and the air, he describes this item as both 'place and space'.[53] Although elsewhere[54] Aristotle does not distinguish between the meanings of *topos* and *chōra* as other philo-

[48] Cf. 208^b6–8: 'for water was formerly in that in which air now is, so that it is clear that place, and the space into which and out of which they moved in moving about, must be something other than either'.

[49] 208^b6: *ἐν ᾧ γὰρ ἀὴρ ἔστι νῦν, ὕδωρ ἐν τούτῳ πρότερον ἦν.*

[50] Aristotle could have produced the following argument: x is somewhere; but in that case where is x? Not in itself (that would be absurd), therefore in something else; hence places are different from their occupants. Presumably he does not use this argument since someone suspicious of places might attempt to deny that anything is anywhere. Denying that anything ever replaces anything else is more difficult, especially for anyone interested in studying nature.

[51] *Phys.* IV 1, 208^b27–9: *ὅτι μὲν οὖν ἐστί τι ὁ τόπος παρὰ τὰ σώματα . . . διὰ τούτων ἄν τις ὑπολάβοι.*

[52] 208^b3–4: *τὸν αὐτὸν τόπον τοῦτον . . . κατέχει.* For more on this locution see below, Chapter 2, p. 78.

[53] 208^b7: *ὁ τόπος καὶ ἡ χώρα.*

[54] e.g. IV 1, 209^a7–8. In the *De caelo* Aristotle regularly uses *χώρα* in contexts where *τόπος* would do just as well. See Bonitz, s.v. *χώρα*.

sophers did,[55] it is none the less the case that *chōra* drops out during the argument of *Physics* IV 1–5 as Aristotle gets closer to defining his notion of place. Presumably one reason for this is that he wants to mark out his notion of place from that of Plato, whose preferred word, for reasons that we shall see, was *chōra*. In any case, Aristotle uses *chōra* in his discussion of replacement to remind us that the water is ceding—for which the appropriate verb is *chōreō*[56]—to the air, and something is ceding to the water, for it too is going somewhere. There is a going-in and a coming-out (both of which are covered by *chōreō*). Aristotle exploits the apparent etymological connection between *chōra* and *chōreō*.[57]

Replacement is a perfectly ordinary phenomenon: the bottle which was full is now empty; the glass which was empty is now full (in both these cases 'empty' means, according to Aristotle, 'filled with air'). But Aristotle does not wish to restrict himself to cases where the replacement occurs in a vessel, for he considers something going out and something coming in, *as if* from a vessel.[58] There is an analogy between cases of replacement in a vessel, and those where there is no vessel. There is no mystery here—the analogous cases in question are perfectly ordinary cases of *dis*placement. When I dive into the lake, there is an *antimetastasis*; I replace a certain amount of water; I am where that water was. This is not replacement in a *vessel*, for there is no vessel which persists, but rather replacement in a *place* (this is part of what Aristotle means when he makes the simple analogy between places and vessels).[59]

Now, Aristotle believes that the universe is full and of fixed shape, and so he thinks that *all* cases of local motion involve such replacement (for instance, when I move forwards, I replace the air which was in front of me), for he is committed to the view that when something moves, things must reorganize themselves in order to accommodate that movement. As Aristotle says in *Physics* IV 7, motion is possible in a plenum, 'for it is possible for things to make

[55] My translation of the one as 'place' and the other as 'space' is only to signal that two different Greek words are used.

[56] See LSJ s.v. χωρέω.

[57] Thus, we should not confuse the notion of χώρα with the notion of an independently existing *space* or *extension*, against which Aristotle vehemently argues at *Phys.* IV 4, 211^{b}14–29. (I discuss that horribly difficult passage in Chapter 4, 'Space'.) Simplicius exploits the apparent etymological connection between χωρέω, χάος, and χώρα (*In Phys.* 620. 13–15).

[58] 208^{b}2–3: ὥσπερ ἐξ ἀγγείου.

[59] See e.g. IV 2, 209^{b}28–9, and IV 4, 212^{a}14–16.

way for one another simultaneously'.[60] This process is called *antiperistasis* ('mutual replacement')[61] and was the common response of those philosophers who did not believe in the void to the argument that motion implies its existence. Aristotle himself appeals to the rotation of some continuous body, either a solid or a liquid, to show that movement is uncontroversially possible without positing the existence of void—when a top spins round, there is no need to suppose that it contains some void, for all that happens is that its parts simply replace each other simultaneously. This is what happens on a larger scale within the universe: the parts of the universe simply replace one another at the same time. Thus Aristotle draws attention to the phenomenon of replacement, but the example of replacement in a jug is simply an analogy designed to get us to think about locomotion in the right way, for locomotion is simply a question of *replacement in a place*, i.e. *displacement*.

Mutual replacement (*antiperistasis*) can be defined as follows: when a body, M_1, moves, it replaces another body M_2, which replaces another body M_3, . . . which replaces a body M_n, which replaces M_1.[62] It is not clear whether *antimetastasis* ('displacement') must also involve *reciprocal* replacement. In Aristotle's general description of *antimetastasis* in *Physics* IV 1, 208ᵇ1–8, air comes to be where water was, i.e. in a particular vessel, but he does not say whether the water is then where the air was (if it is, then it is presumably only momentarily there). When water runs out of a bath, it is replaced by air, but has there been *antimetastasis* of water by air? Aristotle talks of the case where *x* replaces *y* and *y* replaces *x* by saying that *x* and *y* replace *one another*,[63] but if *antimetastasis* always involved mutual replacement this would be pleonastic. Thus it seems that *antimetastasis* need not be mutual. In any case,

[60] IV 7, 214ᵃ29–30: *ἅμα γὰρ ἐνδέχεται ὑπεξιέναι ἀλλήλοις.*

[61] See Barnes [27], 399–402. (Barnes translates *ἀντιπερίστασις* by 'counter-circulation'.)

[62] This is exactly how Simplicius defines *ἀντιπερίστασις*: *In Phys.* 1350. 31–6: *ἀντιπερίστασις δέ ἐστιν, ὅταν ἐξωθουμένου τινὸς σώματος ὑπὸ σώματος ἀνταλλαγὴ γένηται τῶν τόπων, καὶ τὸ μὲν ἐξωθῆσαν ἐν τῷ τοῦ ἐξωθηθέντος στῇ τόπῳ, τὸ δὲ ἐξωθηθὲν τὸ προσεχὲς ἐξωθῇ καὶ ἐκεῖνο τὸ ἐχόμενον, ὅταν πλείονα ᾖ, ἕως ἂν τὸ ἔσχατον ἐν τῷ τόπῳ γένηται τοῦ πρώτου ἐξωθήσαντος* ('There is mutual replacement whenever some body is pushed out by some body and there is an exchange of places; the pusher stands in the place of the pushed and the pushed pushes the next one, i.e. that with which it is in contact, if there happen to be any more of them, until the last comes to be in the place of the first which was pushed').

[63] e.g. *Phys.* IV 2, 209ᵇ25: *ἀντιμεθισταμένων ἀλλήλοις.*

what is important is that, even if one thing replaces another, and that second thing does not itself replace the first, none the less *something* must replace the first thing.[64]

The point to retain is that although Aristotle establishes that places exist by appealing to a simple case of *antimetastasis*, there is an implicit appeal to the facts of motion more generally, since motion is nothing other than replacement in a place. Something moves by replacing something else, but replacements of this sort show that the place of something cannot be identified with that thing; things are somewhere, and the item which you will mention when you say where something is will be a different item from the thing itself. Places exist.

The existence of places (2): natural motion and the elements

Aristotle's second prima facie consideration for the existence of places starts as follows:[65] 'Again, the local motions of the natural simple bodies (such as fire and earth and the like) not only show that place is something, but also that it has a certain power. For each [natural simple body], if not stopped, moves to its own place, either above or below.' The local motions of the natural simple bodies are the *natural* motions of the elements ('natural simple bodies'). When the elements move naturally, they go in certain fixed directions; some go up and some go down. According to Aristotle, these natural motions are due to the fact that the elements have the tendency to go to their own places—broadly speaking, the elements go to these places unless they are stopped from doing so. Obviously, Aristotle in this passage does not give an *argument* for the doctrine—known as the doctrine of 'natural places'[66]—that the proper explanation of the movement of the elements is that they have their own places to

[64] The two notions of *ἀντιπερίστασις* and *ἀντιμετάστασις* seem to have been linked by Strato of Lampsacus, Theophrastus' successor as head of the Peripatos, whom Simplicius reports as having given an example of *ἀντιπερίστασις*: *In Phys.* 659. 23–6 = fr. 63. 27–30 Wehrli: *ἐὰν γὰρ εἰς ἀγγεῖόν τις πεπληρωμένον ὕδατος ψηφῖδα ἐμβαλὼν καταστρέψῃ τὸ ἀγγεῖον ἐπὶ στόμα ἐπέχων τὴν ἔκροιαν, ἡ ψηφὶς ἐπὶ τὸ στόμα τοῦ ἀγγείου φέρεται ἀντιμεθισταμένου τοῦ ὕδατος εἰς τὸν τῆς ψήφου τόπον* ('if you put a pebble into a vessel filled with water and then invert the vessel while holding the stopper over its mouth, the pebble will move to the mouth of the vessel *as the water exchanges places to take up the place of the pebble*').

[65] 208^b8–12.

[66] This appellation is unfortunate and non-Aristotelian: see below, p. 34.

which they tend to go; he simply uses it as a consideration which weighs in favour of places. For clearly, if the elements move to their own places, then places exist ('place is something'), and in accounting for the movement of the elements, they make some sort of difference to the world (this seems to be part of what Aristotle means when he says that they have 'a certain power').[67]

The doctrine of 'natural places' is a scientific hypothesis, based on and intended to explain observed facts. So in the *De caelo* Aristotle uses the doctrine to explain why the elements move as they do, but also why the earth is at the centre of the universe and is spherical,[68] why things are heavy or light,[69] etc. As such, the doctrine itself is not *quite* the right subject-matter for the *Physics*, which aims to *analyse* the basic notions used in the study of nature, such as cause, change, motion, continuity, etc. It is, however, ideal as a *datum* for such analyses, and is used as such in the analysis of place (here and importantly in IV 4 and 5). It is also an important ally in the arguments against infinite body[70] and the void,[71] and when Aristotle argues that there are in fact things at rest (viz. those things in their 'natural' places).[72]

Since the doctrine is supposed to account, first of all, for the natural motions of the elements, it is important to get clear about (1) what an element is; (2) what a natural motion is; (3) what the differences between the natural motions of the elements are.

(1) The four elements which Aristotle has in mind here are earth, water, air, and fire, out of which all animate and inanimate bodies in the sublunary world are made, in differing proportions and densities.[73] (This is why the elements are called 'simple'[74]—because everything else is made of them.) An Aristotelian element seems to be a scattered body,[75] so the element water, for instance, is the sum of all the water in the world. Things made entirely out of water, such as rivers, are parts of the element water, as is the water present in things not made entirely out of water (e.g. the water in mud).

67 See below, 'The power of place'.

68 *DC* II 14.

69 *DC* IV 3.

70 *Phys.* III 5, 205^a8–206^a8 (cf. *DC* I 6, 273^a7–21, etc.).

71 *Phys.* IV 8, 215^a1–14. See Makin [120], 105–15, for discussion.

72 *Phys.* VIII 3.

73 Perhaps Aristotle leaves the fifth element out of the discussion here because its natural movement—circular motion—does not make sufficiently evident the existence of places: after all, the heavens *rotate* but do not have a *place*.

74 208^b9: ἁπλῶν.

75 See Quine [126], §§ 19–20, for the view that things like water etc. are scattered objects.

(2) Aristotle calls the elements 'natural': something is natural if it contains a principle of change and rest, because of *what it is*.[76] The change in question for the elements is locomotion, i.e. change of place. The elements can initiate their own changes of place. For instance, when you drop a clod of earth it falls, without having to be pushed by something else. When you kindle a fire, the flames go up, without having to be forced upwards by something else. To each element corresponds a particular motion which it can initiate, and hence these motions are natural—natural *for* whichever element is in question. The elements move naturally but not in the same way as animals, because when an animal moves, one part of it moves another part of it; the elements, on the other hand, are undifferentiated simple bodies, and hence one part does not move another part. This is to be expected, because the elements are not animate.[77] The movements of the elements are therefore natural in that all that needs to happen is that the external conditions are appropriate, and then they will move (this is captured in our passage by the phrase 'if it is not stopped from doing so'). The elements do not *need* to be given a push, or have some external force intervene in order to get them moving; it is enough that they are not prevented from moving.[78]

Co-ordinate with natural motion is natural rest:[79]

> All bodies both rest and move by force and naturally. A body moves naturally to that place where it rests without force, and rests without force in that place to which it naturally moves. It moves by force to that place in which it rests by force, and rests by force in that place to which it moves by force.

Something is therefore *of a nature* to be somewhere, when it remains there naturally.[80] Hence the elements have in them a principle of motion and rest, for they move naturally, and stop naturally. So a

[76] See above, 'The importance of motion'.

[77] See *Phys.* VIII 4, 255a5–18.

[78] Aristotle muses on this further in *Phys.* VIII 4, 254b33–255b31 (and there is a back reference to this discussion at *DC* IV 3, 311a9–12). There he attempts to show that although the elements move naturally, none the less they do not contradict the principle that everything which moves is moved by something. He claims that in fact inanimate bodies have in them a principle of being moved (and—presumably—being stopped), but he does not say what actually moves them, although he offers two possible *incidental* movers, namely that which removes whatever obstacle is in their way, or that which brought them into being.

[79] *DC* I 8, 276a22–6. Cf. also *DC* III 2.

[80] e.g. *Phys.* III 5, 205b4: *πεφυκὸς ἐνταῦθα εἶναι*; 205b5–6: *οὗ πεφυκός*.

statement like 'earth moves naturally downwards' does not mean that *in all circumstances* earth tends to move downwards, for when earth is naturally at rest, it has no tendency to move at all.

Aristotle more often than not talks of the natural movement of an *element*, but we are more obviously acquainted with—and in physics seek to explain—the natural movements of those things *made out of* the elements. Aristotle thinks that lumps of earth and other objects made of earth (e.g. stones) have the tendency to move downwards, as well as the whole element earth; similarly, sparks fly upwards, as does fire. If we were to look for a simple characterization of the relation between the movement of an element and the movement of those things made of that element (i.e. the elements' parts), we would presumably just say that the movement of the element as a whole is the sum of the movements of its parts. But it is the natural locomotion of the element as a whole which *explains* the natural movement of anything made from the element, for the proportions in which the elements are present in some composite inanimate body determine what its natural motion is,[81] so a clod of earth moves naturally in a certain way (or is naturally at rest) *because* it is made of earth. One of the challenges of Aristotle's physical theory is to state exactly the relation between the natural motion of an element E and that of something made of E, and this is something which I attempt below.

(3) The different natural motions which correspond to each element actually play a role in *distinguishing* the elements from each other because the elements are what they are partly in virtue of the fact that they move naturally in particular ways: earth would not be earth if it did not tend downwards.[82] Now, earth tends downwards and fire tends upwards, but things are less simple with water and air. Sometimes Aristotle tells us that the natural movement of air is also upwards, and the natural movement of water is also downwards.[83] However, at other times he says that air and water *sometimes* move naturally up, and *sometimes* down.[84] Water sometimes tumbles down naturally (e.g. when it rains), but sometimes wells up naturally (e.g. in springs and sources). Similarly, air rises

[81] For an example see *Meteor*. IV 7, 383^{b}20–6, where the fact that olive oil floats on water is attributed to the fact that, apart from earth and water, it contains lots of air. For a more general statement of the principle see *DC* IV 4, 311^{a}30–3.

[82] See below, 'The power of place'.

[83] Cf. e.g. *DC* I 2, 269^{a}17–18.

[84] Cf. e.g. *DC* IV 3, 310^{a}17–18: *τὰ μὲν ἄνω φέρεται τὰ δὲ κάτω τῶν σωμάτων ἀεὶ κατὰ φύσιν, τὰ δὲ καὶ ἄνω καὶ κάτω.*

in water (e.g. when air bubbles go up), but is displaced downwards by fire (and hence hot air rises above cold air).

To capture these facts, Aristotle uses the notions of weight and lightness, which he ties to the tendencies to go up and down.[85] He actually defines lightness and heaviness this way: 'the heavy is that which goes towards the centre [of the universe], and the light is that which is that which goes away from the centre'.[86] The basic notions are in fact to be *heavier than* something and to be *lighter than* something. Earth is *absolutely* heavy, and fire *absolutely* light—that is, earth is heavier than every other element, whereas fire is lighter than every other element. However, air and water are *relatively* light and *relatively* heavy respectively. Air is heavier than fire, but lighter than water and earth. Water is heavier than fire and air, but lighter than earth. So water has the tendency to rise above earth but sink below fire and air, whereas air has the tendency to rise above earth and water, but fall below fire.[87] Aristotle has in mind facts like the following: when a cavity fills with water, the water displaces the air up and out of the cavity and replaces it; when a stone is placed on the surface of water, it sinks to the bottom of the water and displaces the water upwards. The importance for Aristotle of replacement and displacement in the study of motion is once again evident; there is an obvious connection between the theory of natural *motion* and a theory of natural *displacement*.[88] Although Aristotle couches these observations in terms of the behaviour of the elements, the propositions hold true for inanimate bodies made up of the elements, in other words, stones, rivers, pockets of air, flames, etc.[89]

Aristotle *explains* the natural movements of the elements—in-

[85] There are certain problems with Aristotle's account of heaviness and lightness, in particular his contention that the elements (except fire) have weight even in their own places. In their own places the elements have no tendency to go down, because they are naturally at rest. However, these problems can be left to one side for the purposes of this discussion.

[86] *DC* I 3, 269^b23–4.

[87] See *DC* IV 4 for the details of this.

[88] Cf. *Phys.* IV 8, 216^a29–33: *καὶ αἰεὶ δὴ ἐν παντὶ σώματι ἔχοντι μετάστασιν, ἐφ' ὃ πέφυκε μεθίστασθαι, ἀνάγκη, ἂν μὴ συμπιλῆται, μεθίστασθαι ἢ κάτω αἰεί, εἰ κάτω ἡ φορὰ ὥσπερ γῆς, ἢ ἄνω, εἰ πῦρ, ἢ ἐπ' ἄμφω, [ἢ] ὁποῖον ἄν τι ᾖ τὸ ἐντιθέμενον* ('This is always the case for every body that admits of change of place: that if it is not compressed it changes place, to the extent that its nature allows—either always downwards, if its motion, like that of earth, is downwards, or upwards, if it is fire, or either way—whatever the inserted object may be').

[89] The theory of relative weight and the tendencies that the elements have to order themselves is not unlike a theory of specific gravity, as Hussey points out ([5], 223):

cluding these tendencies to rise or sink one above or below the other—in the following way. He imagines what the sublunary world would look like if all the elements were to arrange themselves as they should, i.e. if the elements fulfilled all their natural tendencies to move in their various directions in their various media. Then, there would be a sphere of earth at the centre of the universe, surrounded by a spherical layer of water, which would in turn be surrounded by a spherical layer of air, which would in turn be surrounded by a spherical layer of fire. These four 'concentric spheres' or layers of body would be in four different places (on any account of what a place is). Call these respectively the place of earth, the place of water, the place of air, and the place of fire. Aristotle's explanation for the natural movements of the elements is that each is actually naturally going to its corresponding place or 'its *own* place', as he says in *Physics* IV 1.[90] These places are where the elements are of a nature to be.

This theory attempts to account for the elements' being disposed as they are in actuality. The element earth, for example, has the tendency to go down towards the centre of the universe, and the tendency to collect *around* the centre of the universe and to 'even itself out' around that point. Hence the earth (what we would call the *planet* earth) is in fact more or less spherical and at the centre of the universe because when in its own place, earth would be arranged as a perfect sphere around the centre of the universe.[91] The same is true of the other elements, thus the fact that the element earth would be spherical explains why the other elements arrange themselves in spherical layers around it—water would tend to even itself around earth etc.[92] Once again, this theory is supposed to account not only for the fact that the elements move in these various ways, but also for the fact that they are *actually*—for the most part—arranged in these places.

In terms of the centre and extremity of the universe, if *x* rises above *y* then *x* ends up further from the centre of the universe than *y*, and nearer the extremity, whereas if *x* sinks below *y* then *x* ends up nearer the centre of the universe than *y* and further from the

'we are left, in effect, with sixteen different fundamental specific gravities: those of fire in fire, fire in air, air in fire, etc.'

[90] *Phys.* IV 1, 208^{b}11: *φέρεται γὰρ ἕκαστον εἰς τὸν αὑτοῦ τόπον*. The predicate to supply with *ἕκαστον* is 'natural and simple body'.

[91] *DC* II 14, 297^{a}8–30.

[92] *DC* II 4, 287^{a}30–^{b}4.

extremity (the centre of the universe is, as Aristotle puts it, already 'defined').[93]

The theory is supposed to account not only for the *directions* of the natural movements of the elements, but also for the *destinations* of these movements (i.e. where the elements will naturally come to a stop if they are not impeded beforehand or forced onwards by something else). Earth therefore has the tendency not just to go *towards* its place (*direction*), but also to go *to* its place so that it ends up *in* it (*destination*). Earlier, I translated what Aristotle says in *Physics* IV 1 as follows: 'each [natural and simple body] goes to its own place'. The 'to' is important here, for it means 'to' or 'into' rather than 'towards'[94] (cf. the difference between 'to go to Athens' and 'to go towards Athens'). Aristotle does not just mean the elements go *towards* their own places, but also that they go *to* them.[95]

The doctrine of 'natural places' is therefore supposed to account for the natural direction of the elements' natural movements, but also for the natural destination of those movements. Aristotle is trying to explain why it is that fire always goes in the same direction when it moves naturally, and why earth does too.[96] Water and air, however, do not always naturally move in the same direction, for they move naturally both up and down, but they do move naturally towards (and into) their own place, and in *that* sense their movement is always in the same direction.

The elements are broadly speaking in a constant state of flux—there is always *something* going on. How can this be possible if they

[93] *DC* IV 4, 311^{b}21–2: *ὥρισται τὸ μέσον.*

[94] 208^{b}11: *φέρεται γὰρ ἕκαστον εἰς τὸν αὑτοῦ τόπον.* For this meaning of *εἰς* see LSJ s.v.

[95] Aristotle *never* uses *πρός* or *ἐπί* (which are, I take it, words for 'towards') with expressions like *τὸν αὑτοῦ τόπον*, and *always εἰς*. He reserves *πρός* or *ἐπί* (and *ἀπό* for 'away from') for use with expressions for the centre and extremities of the universe, and this is no accident. The centre and the extremities are not the places of the elements.

[96] Notice that for Aristotle two things can go in the same direction, e.g. downwards, but not be tracing paths which are parallel. For since earth always heads for the centre of the universe, the movements of its parts are not parallel, but at equal angles to the surface of the earth (namely, 90°): heavy bodies 'do not fall to earth in parallel lines, but always at the same angle to it' (*DC* II 14, 296^{b}19–20: *τὰ φερόμενα βάρη ἐπὶ ταύτην οὐ παρ' ἄλληλα φέρεται ἀλλὰ πρὸς ὁμοίας γωνίας*). Frege [102], §§ 64–8, defined the notion of the direction of a line by saying that two lines have the same direction iff they are parallel. Even allowing for the fact that this does not cover direction in the sense of *motion* in a direction, Frege would presumably not have said that movements along non-parallel lines are in the same direction.

move naturally to their own places, and stay still there naturally? Surely once in its own place, if some earth is to move naturally again, then it will have to get to a point from which it can fall (in other words, it will have to go up), in which case for any movement down there will have been a movement up, and in which case earth would have to move up as often as it does down, yet this is not the case. The answer to this puzzle in the *De generatione et corruptione* is that the elements sometimes turn into one another.[97] So water sometimes turns into air, and then this air will rise up and might condense and become water again, in which case it will fall naturally, but without having actually gone upwards. (This is, for instance, the explanation of rain.)

Aristotle therefore explains the natural movement of earth and things made out of earth as being a tendency to go to its place. Other philosophers had attempted to explain the natural motion of earth by saying that 'like goes unto like', in other words, when I drop some earth it falls to earth in order to join the rest of the element earth. Aristotle firmly rejects this and claims instead that in the case of earth, it seeks to get to its place at the centre of the universe: 'it is not the case that if someone were to displace the earth to where the moon is now, each of its parts would go towards it, but rather to where it is now'.[98]

The explanatory nature of the doctrine also comes out in his discussion of weight, for to say that each element naturally goes to its own place is not to *describe* the phenomenon of relative weight. Not *every* motion which results from the elements' tendency to reorder will be correctly describable—as it were, geometrically describable—as a motion to the place of the element which composes the body in question. For instance, there are springs on top of mountains which have come about because some water has welled up through the earth, and has probably gone further than the place of water the element.[99] However, the elements arrive at the places which befit[100] them *by* ordering themselves according to

[97] See esp. *GC* II 4.

[98] *DC* IV 3, 310^{b}3–5.

[99] It will of course immediately seek the place of water by the shortest route possible—this is the explanation of why rivers start in the mountains. Cf. *Meteor.* I 13, 349^{b}27–350^{a}4.

[100] This locution is found at e.g. *Phys.* IV 5, 212^{b}33: ἐν τῷ οἰκείῳ τόπῳ. The translation 'proper place' should be reserved for ἴδιος τόπος, which is something quite different (see Chapter 2 below). There is no nuance between talking of an element's own place and the place which befits it.

their relative weights—this is at the same time *how* and *why* they do so. This might sometimes result in motion which actually distances the element in question from its own place, but for the most part, the movements of bodies will be towards the place of the relevant element. The fact that this happens for the most part is all that is necessary for explanations in physics.

It remains to explain how the natural motions of the elements relate to the natural motions of those bodies made from the elements. A first approximation is that a body B made of element E moves naturally to the place of E. This, however, does not account for a certain type of natural movement which B might undergo, namely that *within* the place of E. For instance, a clod of earth when outside the place of earth moves towards it (i.e. downwards), but when it arrives at the place of earth, it does not just stop, for this might actually be in mid-air, if the sphere of earth does not happen at that point to contain some earth! The clod will continue—surely perfectly naturally—*until there is nothing lighter than it below it*. In general, the rule for a body B made from an element E seems to be: B goes naturally as far as it can *into* the place of E, moving naturally up if there is something above it heavier than it which will have the tendency to displace it upwards, or naturally down if there is something below it lighter than it which will have the tendency to rise and displace it downwards.

Does such a body B have its own place, or a place which befits it? Each *element* has just one place which is its own (where it would be if all the elements were in their proper arrangement), and this place stays the same all the time: the place of earth is always in a sphere around the centre of the universe, the place of water is always a spherical shell around that, etc. But there is no exactly corresponding notion of something's own place for *bodies* made of the elements. A stone, for example, does not have its own place to which it tends. For take two stones, one (s_1) next door to the other (s_2), embedded in the earth. Both are naturally at rest. Suppose now you swap them around. Both are still naturally at rest. But in that case s_1's original place and its new place (s_2's original place) should *both* be s_1's own place, and presumably Aristotle intends something to have just one place which is its own, if its tendency to move is to be explained in terms of *that* place. So we should relativize the notion of something's own place to where it is at any one time:

the place which befits a body B made of element E is at time *t* the nearest place within the place of E to where B is at *t*.

As it stands, this needs to be refined further, for as we saw above, B will not necessarily be naturally at rest in virtue of being in the place of E, since it needs to have nothing above it heavier than it and nothing below it lighter than it. But however the account should be refined, it is clear that the place which befits B at time *t* will be *somewhere* within the place of E, determined by where B is and where all other bodies are at *t*. The place which befits B at any one time, and therefore the natural movement of B at any one time, can only be elucidated by reference to the *larger* place which is the place of the element out of which B is made.[101] It makes perfect sense to talk about the place which befits B as being, *broadly speaking*, that of E, but *more precisely*, somewhere or other in that place. This distinction between broad and precise place is characteristic of Aristotle's concept of place, and will return often in what follows;[102] it is important to see that it is latent in Aristotle's doctrine of 'natural places'.

In fact—and despite the impression given by the secondary literature—Aristotle nowhere talks of 'natural' places.[103] It is easy to see why. There is no special *kind* of place, viz. *natural* ones, in the way that there are (for instance) bodies which are natural or unnatural. Places might be called natural or unnatural *for* particular bodies, and indeed Aristotle does talk of places which belong 'by nature' to such-and-such an element.[104] 'Natural' can qualify 'motion' or 'rest', but not 'place'. For this reason, I have been careful to use quotation-marks when talking of 'natural' places.

Aristotle supposes that the explanation of the natural movements of the elements is that they are heading for particular *places*; they are going *somewhere*. This theory is supposed to explain in an obvious way why the elements move as they do. Rather less obviously, it is supposed to explain why mountains are prone to avalanches and subsidence, and tall buildings are more unstable than low ones—it is because the elements tend to 'even themselves out' in their places.

[101] Aristotle is getting at this at *Phys.* III 5, 205^{b}20–2: *τοῦ γὰρ ὅλου καὶ τοῦ μέρους ὁμοειδεῖς οἱ τόποι, οἷον ὅλης γῆς καὶ βώλου κάτω καὶ παντὸς πυρὸς καὶ σπινθῆρος ἄνω.* Cf. also *DC* II 13, 295^{b}21–2: *ὅπου γὰρ ὁτιοῦν φέρεται μέρος αὐτῆς, ἀναγκαῖον ἐνταῦθα φέρεσθαι καὶ τὴν ὅλην.*

[102] See Introduction, pp. 6–9; Chapter 2, pp. 58–66; Chapter 5, pp. 133–4, and *passim*.

[103] Assuming that this would correspond to *φυσικοί*.

[104] *DC* I 9, 278^{b}30–4.

Moreover, it is supposed to explain the phenomenon of relative weight, that is, why hot air rises (the mix of fire and air yields something lighter than air), why air bubbles go up in water, why flames go up. Small wonder, then, that Aristotle should feel sufficiently confident to appeal to his theory for prima facie evidence that places exist: just as the phenomenon of replacement could not be elucidated without using the notion of something's being now *where* something else was, so the natural movements of the elements cannot be properly explained without appealing to *where* inanimate physical bodies go and come to rest naturally.[105]

The six dimensions (1): preliminaries

In the last section I tried to give some background to Aristotle's simple assertion that the natural simple bodies go to their own places when nothing hinders them, some going up, others going down. Aristotle uses the adverbs *anō* and *katō* to mean 'up' and 'down'. But he continues by saying that *to anō* and *to katō* 'and the remainder of the six dimensions' are 'the parts and kinds of place'.[106] The other four dimensions (*diastaseis*) are ahead, behind, right, and left. I propose to spend some time now investigating the question of what a dimension is. Since they are the 'parts and kinds' of place, it is not unreasonable to hope that this will shed some light on Aristotle's concept of place.

The parts and kinds of place are its species (elsewhere Aristotle refers to the dimensions as being the *differentiae* of place,[107] which confirms this), and the sense in which the dimensions are the species of place is elucidated by a passage in *Physics* III 5:[108] 'that which is in a place is so because it is somewhere, and that is either above or below or in some other of the six dimensions'. So anything which

[105] This argument cannot show—and is not meant to show—that places are separate items in the ontology from physical bodies, just that some notion of a place is used in the explanation of the movements of natural bodies. But put this argument alongside the argument which shows that a place is different from what occupies it, and then we are beginning to see that places 'take on' a life of their own (although one which may well be perfectly explicable in terms of the features of physical bodies).

[106] *Phys.* IV 1, 208^{b}12–13: *ταῦτα δ' ἐστὶ τόπου μέρη καὶ εἴδη, τό τε ἄνω καὶ τὸ κάτω καὶ αἱ λοιπαὶ τῶν ἕξ διαστάσεων.*

[107] *Phys.* III 5, 205^{b}32: *διαφοραί.*

[108] 206^{a}5–6: *οὕτω καὶ τὸ ἐν τόπῳ ὅτι που, τοῦτο δὲ ἢ ἄνω ἢ κάτω ἢ ἐν ἄλλῃ τινὶ διαστάσει τῶν ἕξ.* I have translated *ἄνω* and *κάτω* here as 'above' and 'below': I discuss this matter in the next paragraphs.

is in a place is in one of the dimensions, just as anything which is an animal belongs to one of the species lion, tiger, leopard, etc.,[109] and a dimension is therefore something *in* which something can be. 'Dimension' is therefore a rather unhappy translation of *diastasis*, since it is obvious that Aristotle does not mean a dimension in the sense in which *we* talk of dimensions, otherwise he would be claiming that everything in a place is in one dimension, i.e. one-dimensional! It is not clear how to translate *diastasis*, nor, as it happens, the expressions which Aristotle uses to refer to the dimensions. The expression *to anō*, for example, is formed from the definite article plus a word which has a variable syntax, and correspondingly different senses—and can be translated as 'above', 'up(wards)', 'upper', or 'on high'. Now, the six dimensions fall easily into three pairs, up and down, ahead and behind, right and left,[110] and in order to unravel the knotty question of the six dimensions, I propose to spend some time looking at the *expressions* which Aristotle uses for each pair.

(1) Up and down (*to anō* and *to katō*). I discern, roughly, four different uses for the words *anō* and *katō*. (*a*) They can be adverbs characterizing the direction of a motion, and in this sense they are best rendered in English by 'up(wards)' and 'down(wards)'. (*b*) They can also be prepositions, in which case they correspond to the English 'above' and 'below'. Notice that corresponding to the difference of syntax between (*a*) and (*b*) is a difference in, as it were, logical grammar: one appears to be relative, whereas the other does not. 'Upwards' comes in on its own to qualify a verb (e.g. 'sparks fly upwards'), whereas the preposition 'above' needs something following it (e.g. 'the moon is above *the earth*'). Being a spatial adverbial expression, 'above *x*' can be used with the copula—as in my example—as well as with other verbs (e.g. 'Trees grow above ground'). (*c*) There are several quasi-adjectival uses. According to one use, *anō* and *katō* correspond to the adjectives 'upper' and 'lower', so *to anō morion* and *to katō morion* mean 'the upper part' and 'the lower part', and sometimes you find just *to anō* and *to katō* used to refer to the upper and lower parts of something, including animals (this use is frequent in Aristotle's biological works).[111] They

[109] Things are a little more complicated than this, and I return to the question at the end of this section.

[110] Cf. *IA* 2, 704^{b}20–2: *συζυγίαι δὲ τρεῖς, μία μὲν τὸ ἄνω καὶ τὸ κάτω, δευτέρα δὲ τὸ ἔμπροσθεν καὶ τὸ ὄπισθεν, τρίτη δὲ τὸ δεξιὸν καὶ τὸ ἀριστερόν.*

[111] See in particular *IA* 4, *passim*.

can also be used to mean 'upper' or 'lower' in expressions like 'the upper city'. (This is slightly different from the previous use, for 'the upper city' means 'the upper part of the city', not 'the upper of two cities', whereas 'the upper part' means 'the upper of two parts'.) In a third adjectival use, *anō* can be used to mean, simply, 'high up'. In this sense, Aristotle talks of 'the body on high' (i.e. the celestial spheres)[112] and 'the place on high'.[113] Finally—(*d*)—one can simply use the expressions *to anō* and *to katō* on their own, most notably in the expressions *eis to anō* and *eis to katō*, which are simply synonyms for *anō* and *katō* in sense (*a*) and, although this is less common, *en tōi anō*, which is a synonym for *anō* in the sense 'on high'.[114]

(2) Ahead and behind (*to (em)prosthen*[115] and *to opisthen*). *Prosthen* and *opisthen* are used in many ways, mostly corresponding to the uses of *anō* and *katō*. You find the adverbial 'directional' sense (*a*) of each, viz. 'forwards' and 'backwards', as well as the prepositional use (*b*) of each, viz. 'in front of' (i.e. 'before' in the spatial sense) and 'behind'. Equally, you can use the words to mean 'fore' and 'hind', corresponding to use (*c*). As with *anō* and *katō*, this use is common in Aristotle's biology: 'hind quarters' and 'fore quarters' in Greek are *ta opisthen* and *ta emprosthen*.[116] But again—use (*d*)—a synonym for *prosthen* in the adverbial, directional, sense is *eis to prosthen* (*mutatis mutandis* for *opisthen*).

(3) Finally, there is left and right. Things are simpler here, because 'left' and 'right' in Greek as in English are simply adjectives: *aristeros* and *dexios* respectively. The biological use is clear: *to dexion* can mean 'the right-hand side' or 'right-hand part' of something. (The same goes, obviously, for *aristeros*.) However, in order to obtain the adverbial, directional use, you have to construct a complex expression such as *eis to dexion*, or *epi ta dexia* or something like that (just as in English you can say 'to the right'). Notice, however,

112 *DA* II 7, 418^{b}9, 12: τὸ ἄνω σῶμα.

113 *DC* II 5, 288^{a}4 etc.; *Meteor.* I 3, 339^{b}37 etc.

114 The various different uses are obviously related. As an indication of this, consider how the *OED* defines 'up' as 'to or towards a point or place higher than another and lying directly *above* it', and 'upper' (in one sense) as 'situated *above* another part'.

115 Aristotle uses ἔμπροσθεν and πρόσθεν indiscriminately.

116 Cf. Bonitz s.v. ὄπισθεν: 'frequens usus voc ὄπισθεν in describendis corporis animalium partibus', with references. There are also adjectives for 'front' and 'back', namely πρόσθιος and ὀπίσθιος.

that this use of *to dexion* (and therefore *to aristeron*) corresponds to use (*d*) of the other two pairs.

The two most important uses for our purposes are uses (*c*) and (*d*). Use (*c*) is important in biology, because it enables us to talk of the top, bottom, front, back, left, and right of something—its dimensional parts, as we might say.[117] Use (*d*) is important because the adverbial—directional—use (*a*) appears to be definable in terms of it. Aristotle was aware of this: he thinks that motion in a particular direction is to be defined as being motion to a particular place. In *De caelo* 1 4 Aristotle says 'the contraries of place are the above and the below, the front and the back, and the right and the left; and the contraries of motion correspond to those of place';[118] what he means is more explicit in *Physics* VIII 8: 'In terms of place, motion upwards is contrary to motion downwards, and motion to the fore is contrary to motion to the back, and motion to the left is contrary to motion to the right: for these are the contraries of place.'[119] The places named by 'the right', 'the left', etc. are contrary one to another, *therefore* motions to these places are contrary one to another.

So it is that *to anō*, *to katō*, *to prosthen*, *to opisthen*, *to dexion*, and *to aristeron* name places,[120] as when we say that something is *on my right* or *left* (i.e. on the right or left of me), or to *the front* or *back* of me ('to the above of me' and 'to the below of me' do not work in English). Aristotle is aware of this 'relative' use of these expressions:[121] 'they [sc. the six dimensions] are not always the same for us, but come to be in relation to our position, according as we turn ourselves about (which is why, often, right and left are the same, and above and below, and ahead and behind)'. When I say that something lies to my left (i.e. the left of me), I refer to the place which is the left of me. But when I perform a half-turn, what was my left (in that sense) is now my right—those things which were to my left will be to my right, assuming that they do not

[117] I have modelled this name on Aristotle's expression 'parts according to the dimensions' (*DC* II 2, 285^b33–286^a1).

[118] 271^a26–8: *εἰσι δὲ τόπου ἐναντιότητες τὸ ἄνω καὶ κάτω καὶ τὸ πρόσθιον καὶ ὀπίσθιον καὶ τὸ δεξιὸν καὶ ἀριστερόν, αἱ δὲ τῆς φορᾶς ἐναντιώσεις κατὰ τὰς τῶν τόπων εἰσὶν ἐναντιώσεις.*

[119] 261^b34–6: *ἐναντία γὰρ κατὰ τόπον ἡ* [sc. *κίνησις*] *ἄνω τῇ κάτω καὶ ἡ εἰς τὸ πρόσθεν τῇ εἰς τοὔπισθεν καὶ ἡ εἰς ἀριστερὰ τῇ εἰς δεξιά· τόπου γὰρ ἐναντιώσεις αὗται.*

[120] This view is not heterodox; cf. Ackrill [83], 97: 'Aristotle thinks of "up" and "down" as naming two places.'

[121] *Phys.* IV 1, 208^b14–18. Translation from Hussey [5], 20–1.

move. Thus, the left of me and the right of me coincide as I turn around. (Equally, what lies on my left now lies in front of me when I turn through 90° to my left, although this result does not appear to interest Aristotle.) At any one moment the six dimensions relative to me will be well defined and distinct, but when I *turn around* they will often change—as Aristotle says, they will not always be the same.[122] The force of 'often' here seems to be that it is none the less sometimes the case when I turn around that these dimensions will remain the same, namely when I turn through 360°.

Aristotle is concerned that this cannot be the sense in which it is correct to say, for instance, that fire goes up, i.e. goes to the place which is above. For what is above me now may be below me, when I change place. Fire, however, always goes to the place which is above, regardless of who is judging or observing. The places sought by these elements—above and below—are always distinct. They can never coincide according to our or anyone else's orientation. If they could, then, for instance, fire would change direction according to where we were, but this Aristotle finds unacceptable. He thinks that there is a way of understanding above, below, etc. in a different way, not relative to us, such that the dimensions are not only relative to us, but also 'distinct and separate *in nature*'.[123] This takes up Plato's claim in the *Timaeus* that 'above' and 'below' are purely relative terms, because what is now called 'above' may be called 'below' later if I were to be at the other side of the globe.[124] Aristotle *accepts* Plato's point, but insists that there is yet *another* way of understanding 'above' and 'below', such that they are independent of us, or whoever is judging:[125] 'Above is not anywhere whatever, but where fire and what is light, move [to]. Likewise, below is not anywhere whatever, but

[122] It is true that as I turn around, what was in front of me will lie behind me, and that what was to the right of me will lie to the left of me. But in order that what is above me become below me, it is not a question of turning around—for the roof which is now above me to be below me, I need to *climb up* onto the tiles. There is no puzzle, however, in what Aristotle says: he is thinking of walking around the earth in a circle to the other side, as shown by his echo at 208^b17–18 (*διὸ καὶ ταὐτὸ πολλάκις δεξιὸν καὶ ἀριστερὸν καὶ ἄνω καὶ κάτω καὶ πρόσθεν καὶ ὄπισθεν*) of Plato *Tim.* 63 A 3–4: *πολλάκις ἂν στὰς ἀντίπους ταὐτὸν αὐτοῦ κάτω καὶ ἄνω προσείποι.*

[123] 208^b18–19: *ἐν δὲ τῇ φύσει διώρισται χωρὶς ἕκαστον.*

[124] *Tim.* 63 A 2–4. See n. 122 above.

[125] 208^b19–22: *οὐ γὰρ ὅ τι ἔτυχέν ἐστι τὸ ἄνω, ἀλλ' ὅπου φέρεται τὸ πῦρ καὶ τὸ κοῦφον· ὁμοίως δὲ καὶ τὸ κάτω οὐχ ὅ τι ἔτυχεν, ἀλλ' ὅπου τὰ ἔχοντα βάρος καὶ τὰ γεηρά, ὡς οὐ τῇ θέσει διαφέροντα μόνον ἀλλὰ καὶ τῇ δυνάμει.*

where heavy and earth-like things move [to]. So they differ not by position alone but in power too.' Above and below as scientific notions applied to nature cannot simply be identified with the regions which just happen to be above (or below) us as we move around, i.e. which 'come to be in relation to our position'.[126] The individuation of these regions cannot depend on *our*, or anyone else's, position.[127]

The following picture has emerged of what Aristotle is saying. At any one moment there are, relative to me, the six dimensions: my right, my left, etc. These are not the parts of my body, but 'individually centred places', which are referred to in expressions such as 'to my right' etc. None the less, the individuation of these places is *through* the parts of the body in question, since what is on my right is so because it is in a certain relation to the right-hand side of my body. In fact, when I turn around the right-hand side of my body turns with me, and this is why something which *was* on my right may *no longer be* on my right when I turn around. The dimensions relative to us are these individually centred places, and they depend for their individuation on the dimensional parts of the individual on which they are centred. However, Aristotle thinks that there is another way of using 'above', 'below', 'right', 'left', 'front', 'back', this time to pick out not the dimensions centred on us or anything else, but rather what one might tentatively call 'absolute places'. Absolute places are those which are used in scientific explanations of a certain type; when the natural scientist says that fire goes to the place which we have been calling 'above' (i.e. *to anō*), he is not referring to some place centred on someone in particular, but simply to a place which exists in nature, in the absolute. Otherwise, it would be necessary to specify relative to whom fire goes up, and Aristotle clearly does not think that this is either necessary or desirable.[128]

[126] 208^{b}16: κατὰ τὴν θέσιν . . . γίγνεται.

[127] Hussey [5], 20, translates ὅ τι ἔτυχεν (208^{b}19, 20) as 'anything you like', but this carries the slightly misleading implication that Aristotle is telling us that above and below in nature are not *wherever we rule them to be*, which although true, is not quite the point he is trying to make: I think he is saying that above and below in nature are not simply *whatever they happen to be given our position.*

[128] Notice that 'above the earth' will not do either, since according to Aristotle, if the earth were displaced, fire would still go in the same direction, which may no longer actually be upwards from the earth.

The six dimensions (2): the 'dimensions' in nature

But what actually are these absolute places, existing in nature? Let us start with the two examples which Aristotle mentions explicitly in our text, namely above and below. Is above supposed to be just where fire goes, or is it also the place where air goes? If it is just where fire goes, then below must be just where earth goes. But then it will not be true that everything that is somewhere is in one of the six dimensions, since anything in the sphere of water or air—e.g. the majority of water and air—will not be.[129] Hence above and below must include the place of air and water respectively. In any case, in defining *to anō* as where fire goes, Aristotle appears to rule out that the stars, sun, and moon are on high, for they are certainly not where fire goes!

Sometimes Aristotle appears to suggest that *to anō* refers to the outer circumference of the sphere of fire, and *to katō* to the centre of the universe, but this has the unacceptable consequence that everything in the sublunary world is in *to anō*, and raises the interesting question as to how *anything* could be in *to katō*. But perhaps this is being too literal-minded. Aristotle can speak as loosely as he likes. Above is where the spheres of air and fire are, as well as the outer celestial spheres. Below is the where the spheres of earth and water are. Saying that above and below are the outer circumference and centre of the sublunary universe respectively would have to be taken as a loose *façon de parler*—yet another use of the terms *to anō* and *to katō*.

The situation is rather worse when it comes to the other four dimensions. As Hussey [5], 100, says: 'Aristotle's attempt to define an absolute, cosmic sense of the other two pairs is one of the more curious parts of his cosmology.' Aristotle does not say in the *Physics* how to individuate the dimensions right, left, front, and back. Commentators habitually point to *De caelo* II 2, where Aristotle debates the question, following the Pythagoreans, whether there is a right- and left-hand side of the heaven,[130] and inspired by this, whether it also has a front and a back, a top and a bottom—in

[129] Aristotle does refer to a place called 'the middle' at *IA* 5, 706^{b}3, but this really introduces more problems than it solves, not least because it is never referred to as a dimension.

[130] *DC* II 2, 284^{b}6–7: εἶναί τι δεξιὸν καὶ ἀριστερὸν τοῦ οὐρανοῦ.

short, whether the heaven has dimensional parts (a top, bottom, front, back, right, and left). 'The heaven' here means the outer sphere of the heaven (the sphere of the stars), and thus it should be obvious that Aristotle is *not* answering in the *De caelo* the question which arises from the *Physics*. Right, left, front, and back in the *Physics* are dimensional places in nature which are to be distinguished from individually centred dimensional places. However, right, left, front, back, above, and below in the *De caelo* are dimensional parts of the celestial sphere.

Aristotle discerns the dimensional parts of the celestial sphere as follows. First, he refers the reader back to *De incessu animalium* 4, where he had outlined how one might attribute dimensional parts to a body. This can be done by *position* relative to the earth and universe, and by *function*.[131] The differentiation of dimensional parts according to function applies only to living things, for the functions in question are growth, locomotion, and perception. All living things grow, so all living things have an upper and a lower part (*to anō morion* and *to katō morion*) since the upper part is defined as that from which growth and the distribution of food proceeds.[132] Some living things also have the power of perception—this is in fact what distinguishes animals from other living things (e.g. plants)—and the location of the organs of perception is what defines the front and back of an animal.[133] Finally, some (but only some) animals also have the capacity for movement, and it is this which defines which is their right and left side: the part where motion comes from is the right side.[134] Therefore some things have an upper and lower part,

[131] *IA* 4, 705ᵃ31–2: *διείληπται δ' ἔργῳ, καὶ οὐ θέσει μόνον τῇ πρός τε τὴν γῆν καὶ τὸν οὐρανόν.* In *De caelo* Aristotle also distinguishes a *third* sense in which one might attribute dimensional parts to certain (inanimate) things, namely *in relation to us* (II 2, 285ᵃ1–6): *ἀλλ' ἐν μὲν τούτοις λέγομεν τὸ ἄνω καὶ κάτω καὶ τὸ δεξιὸν καὶ ἀριστερὸν πρὸς ἡμᾶς ἐπαναφέροντες· ἢ γὰρ κατὰ τὰ ἡμέτερα δεξιά, ὥσπερ οἱ μάντεις, ἢ καθ' ὁμοιότητα τοῖς ἡμετέροις, οἷον τὰ τοῦ ἀνδριάντος, ἢ τὰ ἐναντίως ἔχοντα τῇ θέσει, δεξιὸν μὲν τὸ κατὰ τὸ ἡμέτερον ἀριστερόν, ἀριστερὸν δὲ τὸ κατὰ τὸ ἡμέτερον δεξιόν* ('It is in relation to ourselves that we speak of above and below, or right and left, in these objects. We name them either as they correspond with our own right hands (as in augury), or by analogy with our own right hands (as with the right-hand side of a statue), or from the parts which lie on opposite sides to our own, i.e. calling right in the object that part which is on our left-hand side, and vice versa').

[132] *IA* 4, 705ᵃ32–3: *ὅθεν μὲν γὰρ ἡ τῆς τροφῆς διάδοσις καὶ ἡ αὔξησις ἑκάστοις, ἄνω τοῦτ' ἐστίν.*

[133] *IA* 4, 705ᵇ10–11: *αἴσθησιν γὰρ ἔχει ταῦτα πάντα, ὁρίζεται δὲ κατὰ ταύτην τό τε ἔμπροσθεν καὶ τὸ ὄπισθεν.*

[134] *IA* 4, 705ᵇ18–20: *ὅθεν μὲν γάρ ἐστι τοῦ σώματος ἡ τῆς κατὰ τόπον μεταβολῆς ἀρχὴ φύσει, τοῦτο μὲν δεξιὸν ἑκάστῳ.*

some of which in turn have a front and back, some of which in turn have a right and left.

Only a very restricted set of things have a right and a left side defined by function, and this is the reason why Aristotle raises the question in the *De caelo* of *whether* the celestial sphere has a right and a left side. It contains a source of movement, i.e. it is a natural body, and is alive (or can at least be fruitfully considered to be alive),[135] so Aristotle can embark on individuating its 'parts according to the dimensions, i.e. those defined according to place'.[136] The obvious problem, however, is that the universe neither grows nor perceives, and hence apparently cannot be said to have an above or a below, nor a front and back (at least following the guidelines of *De incessu animalium*). However, since the celestial sphere moves and has a soul, and is therefore like a mobile animal, it *must* have a right and left—and therefore *must* have an above, below, front, and back (since anything with a right and left defined by function has an above, below, front, and back). As Aristotle had said in the *De incessu animalium*, the right-hand side is that from which motion starts, and so the right-hand side of the celestial sphere is that from where the stars rise (and the left that side where the stars set).[137] Aristotle then argues (rather obscurely) that the invisible celestial pole is the upper pole, and the north celestial pole is the lower, and so the cut-off point between the upper and lower parts of the celestial sphere is the great circle perpendicular to the axis of rotation of the sphere, the *lower* hemispherical shell being that which contains the *northern* celestial pole, and the *upper* one being the one which contains the *southern* (invisible) celestial pole. Similarly, the hemispherical shell through which the stars travel between rising and setting is the *front* one, and the one opposite to it is the *back* one.[138]

Of course, the definition of above and below in the *Physics*—where fire and light things go, and where earth and heavy things go, respectively—does not match up with this at all. As Ross puts it in his note on *Physics* IV 1, 208^{b}13–14:[139]

135 See *DC* II 2, 285^{a}29–30: *ὁ δ' οὐρανὸς ἔμψυχος καὶ ἔχει κινήσεως ἀρχήν.*

136 *DC* II 2, 285^{b}33–286^{a}1: *τῶν κατὰ τὰς διαστάσεις μορίων καὶ τῶν κατὰ τόπον ὡρισμένων.* The reference to place here is presumably a reference to the fact that the dimensions are the species of place.

137 *DC* II 2, 285^{b}17–19: *τοῦ δ' οὐρανοῦ ἀρχὴν τῆς περιφορᾶς, ὅθεν αἱ ἀνατολαὶ τῶν ἄστρων, ὥστε τοῦτ' ἂν εἴη δεξιόν, οὗ δ' αἱ δύσεις, ἀριστερόν.*

138 The best commentary on this is still Heath [23], 231–2.

139 Ross [3], 563.

> It is clear that, with Aristotle's conception of the universe as forming a sphere, if up and down, right and left, front and back are to be treated as the six divisions or directions, each pair must be interpreted as being the hemispheres into which the sphere is divided by a plane, and the three planes must be at right angles to each other. This gives quite a different interpretation to 'up' and 'down' from that offered in the present chapter, where 'up' refers to the circumference of the universe and 'down' to its centre. The other interpretation is given to 'up' and 'down' in *De Caelo* ii. 2, where it is argued that the south is the true upper pole of the universe.

But there was never any hope that the two accounts match up, since they deal with different questions. *De caelo* II 2 is concerned with the dimensional parts of the celestial sphere, whereas *Physics* IV 1 is concerned with dimensional places in nature. No doubt Ross was misled by Aristotle's talking of 'the universe' in *De caelo* II 2, and he thought that this was a reference to the whole universe (this is certainly suggested by the wording of his note). However, Aristotle is certainly talking of the celestial sphere in *De caelo* II 2, since the *entire* universe is *unmoving*, and yet Aristotle is talking of something which moves naturally.

There remain two large problems. (1) The celestial sphere is a sphere which turns on its axis and has stars fixed in it. Imagine a star which is now rising. If Aristotle wishes to say that it is in the right-hand side of the celestial sphere, he has to accept that it will *always* be in the right-hand side of the sphere, since the sphere turns on its axis, taking the star with it (the right-hand side of the celestial sphere turns with it, much as my right-hand side turns with me when I turn round). It will follow, unacceptably, that *that* star, and every other star which rises, is *always* in the right-hand side of the celestial sphere. It will also follow, by a similar argument, that any star which sets is *always* in the left-hand side of the celestial sphere. But since every star which rises also sets, this is not very satisfactory. (2) At any one time, a star might be setting in Athens, but rising somewhere else (Aristotle knew this, and it is a simple consequence of his cosmology); hence what is the right-hand side of the celestial sphere depends on where you are observing.

The problem is that a star is rising or setting according to the orientation of this sphere and there is no *part* of the celestial sphere which is where the stars rise. Therefore, given Aristotle's failure to individuate right and left sides to the celestial sphere, perhaps it is possible to adapt his argument to the case which is pertinent to the *Physics*. Presumably, Aristotle will say that a star is 'on the right' (where this is construed as being the dimensional place in nature) when it is rising, and 'on the left' when it is setting. Suppose the universe, i.e. the *entire* universe, to be divided down the middle along a plane intersecting the axis of rotation of the celestial sphere, yielding two hemispheres such that when stars are in one hemisphere they are rising (this will be the right-hand hemisphere) and when they are in the other they are setting (this will be the left-hand one). These two enormous hemispheres are immobile, for the whole universe is immobile, the celestial sphere being merely its rotating outer shell. In other words, it is possible to use the behaviour of the stars to fix dimensional parts of the universe *as a whole*, rather than those of the celestial sphere. This does seem to be what Aristotle has in mind sometimes when he talks of the dimensions in nature as being dimensions 'in the whole itself'.[140]

This attractive version of the absolute places 'right' and 'left' has none the less several major drawbacks. The right- and left-hand sides of the universe will not be the same for two different places, for the simple reason, noted above, that at any one time one and the same star might be rising somewhere and setting somewhere else. The only solution to this conundrum is to suppose that the left- and right-hand sides of the universe are relative to different places, then right and left in the universe will be very similar to what we would more naturally call eastern and western hemispheres of the universe (in fact, it seems to me very probable that Aristotle was feeling for this very distinction). When the sun is rising in Oxford it is in the eastern hemisphere of the universe relative to Oxford, but it is setting in Sydney in the western hemisphere relative to Sydney. The two hemispheres of the universe which correspond to right and left *in the universe* will be different in different places, but in each case they will not depend on any observer's *orientation*, which marks the crucial difference between right and left *in nature* or *in the universe*, and right and left *relative to us*. In strict terms, the dividing plane between the right and left sides of the universe for a particular sublunary place will be that plane which intersects the axis of rotation of the celestial sphere and that place.[141] I have no great confidence in this as a solution to Aristotle's ills, but at

[140] *Phys.* III 5, 205^{b}34: ἐν αὐτῷ τῷ ὅλῳ.

[141] This means that at the north and south poles of the earth there would be no right- and left-hand side of the universe, since no stars rise and set at those places.

least this would have been a way in which he could have used the rising and setting of the stars to define a right- and left-hand side of the universe which does not depend on any observer's *orientation*. The point of this is that it means that a physical science, e.g. astronomy, can refer to the right- and left-hand sides of the universe for a given place, without having to specify any facts about an *observer*. This would clearly be very important to Aristotle, since the heavenly bodies quite clearly obey laws of motion independently of any observer's orientation.

Another problem is that if a similar distinction is operated for front and back and above and below, along the lines suggested by *De caelo* II 2, it will once again yield something completely different from *Physics* IV 1 for above and below. But perhaps the worst problem of all is that we are using the notion of function (where motion starts) in order to define not a dimensional part of something, but a dimensional *place*. The universe as an entirety does not move, so there is no reason to think that it has a right and a left, even if these are relative to particular places on earth. This simple fact may be enough simply to rule out the use that *De caelo* II 2 has for interpreting Aristotle's strange remark in *Physics* IV 1 about the dimensional places which exist in nature, not relative to us.

The six dimensions (3): conclusion

Let me sum up the argument so far: there are places which are called right and left, front and back, above and below, which are centred on us. These change according to our orientation. But there are also a right and a left, a front and a back, an above and a below in nature, which are scientific notions and not dependent on our orientation or anyone else's. As I have tried to suggest, Aristotle is feeling for a distinction which operates rather like the points of the compass. The sun rises in the east, regardless of which way *we* are facing. Aristotle not unreasonably thinks that the notions of east and west etc. can be derived from looking at the risings and settings of the stars. He attempts in *De caelo* II 2 to link this to the differentiation of right and left in an animal, by taking the relevant part of the universe as some sort of animal and thinking of its dimensional parts as defined by its characteristic activity; this attempt does not quite succeed.

These 'absolute' dimensions should also enable us to differentiate the dimensional parts of something according to their position relative 'to the earth and universe'. Presumably all bodies have dimensional parts in this sense. The upper part of something is then simply that half of something which is *above* the other half. This will lead in some cases to the upper half as defined by function being different from the upper half as defined relative to the universe. The roots of a plant are, functionally speaking, its upper part, for this is where it takes in nutrition. But relative to the universe, it is the bottom of the plant. Man's functional upper part, on the other hand, is also his upper part relative to the universe.[142] 'Relative to the universe' thus makes sense for 'upper' and 'lower'—it is just a question of 'above' and 'below' in the sense we have seen before. However, once again, it is mysterious how it can be the case that something's right-hand side is right relative to the universe, since it is not obvious where the right is, in that sense, but if I am right that Aristotle has something like the points of the compass in mind, perhaps he has in mind the distinction between the northern, eastern, southern, and western sides of an object.

Aristotle has a bewilderingly complex (and quite possibly inconsistent) view of what right and left are. The dimensions—the distinction between right, left, above, below, front, and back—can be used to differentiate places, and also parts of a body. The distinction as applied to places yields two different sets of places: those which are relative to us (and defined in fact by our dimensional parts), and those which exist in nature (and these are defined by the behaviour of the elements and the celestial bodies).

The parts and kinds of place

How exactly are the dimensions the parts and kinds of place? Presumably, both the dimensions in nature and 'individually centred places' are parts and kinds of place; at least, Aristotle does not say anything to the contrary. The important point in each case is that together, the dimensions exhaust the possibilities of location, so anything which is somewhere has to be in one of the dimensions.

[142] Aristotle feels that this shows that man is the most *honourable* animal.

But there are two difficulties here. First, the dimensions appear to overlap, with the result that something can be in two dimensions at once, but species do not normally in this way share members. Second, it seems to be possible for one and the same thing to straddle two dimensions, yet normally it is not permitted for something to belong partially to one species, partially another.

Both problems stem from taking as the model for the genus–species relation a relation such as that between the genus *animal* and its species. The overlapping of species (the first problem) is impossible in the case of species of animal. However, it is obvious that the *dimensional parts* of something (its right, left, ahead, behind, above, and below) overlap: take the hind left leg of a dog, for instance. Equally, 'individually centred places' overlap: something may lie at once to my left and in front of me. In fact, it seems perfectly admissible that the species of, say, *colour* overlap: two different colour-words may overlap in their extension (e.g. red and brown, or red and orange).

The idea that something might straddle two species (the second problem) likewise seems impossible in the case of the species of animal: belonging partially to one species and partially to another is some sort of anomaly. But the case under consideration is perfectly analogous to that where something is *partially* coloured blue and *partially* coloured red. Just as blue and red are two different species of colour, so right and left are two species of place, yet something can partially belong to one and partially to the other.

There is an interesting passage in the late Byzantine anonymous *Logica et quadrivium*[143] that develops this kind of interpretation. There are the six familiar species of 'where', namely above, below, in front, behind, right, and left. But then there are further 'differences' of (for instance) *above*, namely among the clouds, in the aether, among the stars, under the sky, above the sky.[144] These are 'more particular places';[145] once again, there is a connection between this and broad and precise places.

Aristotle's second argument for the existence of places is thus slightly clearer now. Natural and simple bodies (i.e. the four elements) move up and down. Their propensity to move in these directions defines them as natural bodies, since it is part of their

[143] Ed. J. Heiberg (Copenhagen, 1929).

[144] 12. 6–7: *τοῦ . . . ἄνω πολλαὶ διαφοραί, ἐν νεφέλαις, ἐν αἰθέρι, ἐν ἄστροις, μέχρι τοῦ πόλου, ὑπὲρ τὸν πόλον.*

[145] 12. 8–9: *τοῖς μερικωτέροις τόποις.*

'principle of motion'. But moving in such-and-such a direction introduces the notion of a place, at least inasmuch as we have to refer to these natural and simple bodies as going *somewhere* and coming to a stop *somewhere*. The directions of the motions of the natural and simple bodies are not to be understood as *relative to us*, for they are independent of us, and in nature. So places which are independent of us have to be accounted for in a physical theory, although we do not yet know what sort of item they are.

It is of course true that Aristotle's cosmological system is faulty. That there are spheres for the elements just does not seem to correspond to the facts, and the theory of crystalline spheres carrying stars has not met with much approval from contemporary astronomers. To what extent does Aristotle's argument for the existence of places depend on the spheres? It seems to me that it does not depend on them at all. Spheres or no spheres, the four elements move in particular directions. We are all familiar with (for instance) the fact that stones sink, i.e. go *down*, in water (and that, all other things being equal, they go straight down). We know that air bubbles rise, i.e. go *up*, in water, but that to keep a hot-air balloon *up* you need fire (i.e. you need to heat the air). A study of nature has to account for these things, and that means having something to say about the elements, directions, and therefore places.

The power of place

The point of Aristotle's opening considerations for the existence of places is to try to get us to accept that both our everyday conception of nature and the science of nature include a conception of places. But the argument from the natural movements of the elements and the dimensions also aims to show that places make some sort of difference to the world, that they have some sort of role to play in the world. Aristotle expresses this by saying that they have some sort of power (*dunamis*),[146] and *justifies* it by saying that each element moves to its own place.[147] It is difficult to shake off the impression that Aristotle thinks that an element is *attracted* to its own place, and that in that way its place exercises a sort of power. However, this

[146] 208^b10–11: *ἔχει τινὰ δύναμιν*. Cf. 208^b21–2: *ὡς οὐ τῇ θέσει διαφέροντα μόνον ἀλλὰ καὶ δυνάμει*. We should resist the temptation just to translate *δύναμις* as 'significance'; there seems to be more to say.

[147] Cf. 208^b11: *γάρ*.

is *not* what he says; the notion of attraction is not Aristotelian, and has been smuggled in by commentators.[148] Perhaps they have been influenced by the word Aristotle uses for 'move',[149] which *literally* means 'is carried to', and have supposed that if Aristotle says that the elements are carried to their own places, then he must mean that it is the *places* which do the carrying. In fact, the word for 'move' which Aristotle uses is the normal one for movement which is not voluntary (no one will claim that the elements move *voluntarily*, even if they move *naturally*), and the perfectly ordinary word he had used just before for 'motion'[150] is etymologically related to it, so the mere use of this word as opposed to any other should not be seen as an attempt on Aristotle's part to imply that the elements are moved *by* their places.

What sort of power *does* Aristotle envisage places having? There is no reason to suppose that this asks what actions places perform, nor what they undergo. Aristotle means simply that places make a difference to the world—they enter into an account of why the world is as it is, for the elements are actually (in part) individuated by *where* they go. This comes out neatly in the following passage from the *De caelo*:[151]

> Further, these worlds, being similar in nature to ours, must all be composed of the same bodies as it. Moreover, each of the bodies, fire I mean, and earth and their intermediates, must have the same power as in our world. For if those elements are named homonymously and not in virtue of having the same form as ours, then the whole to which they belong can only be called a world homonymously. Clearly, then, one of the bodies will move naturally away from the centre and another towards the centre, since all fire must have the same form as fire, and so on with each of the others, as the parts of each have the same form in this world.

Aristotle talks of 'the form of fire', which *defines* fire: 'all fire must have the same form as fire'. For anything to count as fire, it must have a certain form. What is this form? Well, as the 'since' towards the end of the passage shows, the form of fire either is, or consists partially in, its being naturally such as to move away from the centre of the universe, to its own place. Thus, fire's own place plays a part in the account, or form, of fire. The place to which fire tends

[148] See e.g. Ross [3], 563: 'The proper place of a body, according to Aristotle, has an actual influence on it; an attractive influence which draws the body to it.'

[149] 208^{b}11: φέρεται.

[150] i.e. φόρα. Cf. 208^{b}8.

[151] *DC* I 8, 276^{a}30–^{b}7. Translation adapted from the Oxford translation [1].

defines what fire is; fire would not be fire if it did not tend towards this place.

According to Aristotle's theory of causation, some aspects of what something does, or what properties it has, are explained by what the thing actually is. This is the so-called 'formal cause'. Aristotle's notion of a cause (*aitia*) is 'that on account of which something holds of something'.[152] Ask yourself *why* x is such-and-such, and the answer may well be 'on account of y'—then y is a cause of x. x and y are related in one of four ways: y might be the matter of x, the form of x, the source of change of x, the goal of x.[153] Fire goes up because it would not be fire if it did not. In other words, fire goes up because of its form.[154] Places are partial formal causes: places appear in the definition of the elements; they are not to be identified with that definition.[155] (There is clearly a sense in which places are also the goals of the elements' motions, although it is not clear whether we should say that places themselves are the goal of the elements, or being in those places.[156]) So places feature in the definitions of the elements, and in that way explain why, for instance, fire goes up.[157]

When Aristotle talks of the power of places in the *Physics*, he has in mind the power that places have to define what the elements are, for the elements would not be what they are if their own places were not related in the way they are. This tells us something about how places play an explanatory role in, or make a difference to, the world. For instance, earth thrown up into the air will come down again (assuming there is nothing stopping it), because that is what earth does; it would not be earth if it did not come down. Similarly, a fire which starts on the fourth floor will, all other things being equal, work itself upwards to the fifth floor and above; it would

[152] A paraphrase of *τὸ διὰ τί περὶ ἕκαστον* (*Phys.* II 3, 194^{b}19).

[153] Cf. *Phys.* II 3, 194^{b}23–195^{a}3 (=*Meta.* Δ 2, 1013^{a}24–^{b}3).

[154] Something like this surely lies behind Aristotle's cryptic comment that 'Motion towards its own place is for each thing motion towards its own form' (*DC* IV 3, 310^{a}33–^{b}1): *τὸ δ' εἰς τὸν αὐτοῦ τόπον ἕκαστον τὸ εἰς τὸ αὐτοῦ εἶδός ἐστι φέρεσθαι.*

[155] But cf. Aristotle's remark that 'the parts of the definition' (*τὰ μέρη τὰ ἐν τῷ λόγῳ*) are also formal causes (*Phys.* II 3, 194^{b}28–9 =*Meta.* Δ 2, 1013^{a}29).

[156] For more on the final cause see Wieland [131] with Schofield [128]. According to Sorabji, Aristotle believes (though never actually says) that places are final causes ([26], 186–7). Note that for things whose essence is to engage in a certain goal-directed activity, final and formal causes coincide (Schofield [128], 38–9).

[157] This interpretation is shared by Furley in [107], 86 n. 46: '[Place] is . . . a factor in the formal cause . . . of the primary bodies'. Note that this explanation of *why* fire goes up is not supposed to be an explanation of *how* it goes up.

not be fire if it did not. Where something is makes a difference to where it will (naturally) go: for instance, water rises up from the earth but falls down from the air. The reason for this is that water just does this; it would not be water if it did not. This defining feature of water can only be expressed by adverting to *where* water goes—hence the power of place.

In fact, Aristotle, at least sometimes, seems to think that the elements *are* powers, not substances.[158] This appears to mean that the reality of the elements consists purely in the *capacities* with which they endow the substances which are made out of them: the fact that, for instance, heavy things go down, and light things go up,[159] or that hot things heat. To say that something is made out of earth just is to say that it has the capacity, or tendency, to fall, when in the air. Where something goes when in a given place actually helps define what it is that thing is made out of, so the power of place to define what the elements are also helps define what *substances* are made of.[160]

Places have power, i.e. causal significance, according to Aristotle. Some commentators are embarrassed by this, apparently because of a certain prevalent conception of causality.[161] In order to convince themselves that Aristotle could not have thought that places have a causal role, they appeal to the puzzle that Aristotle raises in *Physics* IV 1, at 209^{a}18–22, and claim that Aristotle wishes to endorse its aporetic conclusion, that places are not one of the four causes.[162] Aristotle imagines someone asking the question: to which type of cause would place belong? The next sentence reads: 'for none of the four causes applies to it'.[163] The questioner explains (cf. 'for') why it is that the question is difficult: it is because places do not appear to be any of the four causes. Algra points out that Eudemus, in his *Physics*, thought the same thing, and that Simplicius says that this was 'as agreed'—this must mean that Eudemus agrees with *Aristotle*, and hence that Aristotle too thought that places do not

[158] *Meta.* Z 16, 1040^{b}5–8.

[159] See Barnes [86], 45.

[160] Cf. Aristotle's insistence that the motions of the natural *simple* (ἁπλῶν, 208^{b}9) bodies show that places have power.

[161] That places cannot be causes has become a commonplace. 'If places are causally inert, as the prescientific conception has it, then' (Forbes [135], 296). (In truth, I do not know to what 'the prescientific conception' refers.)

[162] I am thinking especially of Algra ([56], 150).

[163] 209^{a}19–20: *οὐδεμία γὰρ αὐτῷ ὑπάρχει αἰτία τῶν τεττάρων.*

appear to be any of the four causes.[164] But this is a red herring: not only is Sorabji correct to say that 'The denial at Aristotle *Phys.* 4. 1, 209a20 that place can serve as any of the four causes, or four modes of explanation, is merely part of a puzzle or *aporia*',[165] but also it is *true* that places are not one of the four causes: places are only *parts of*, only *appear in*, the formal causes of the elements. (Hence why the puzzle arises.) They have causal significance, but only up to a certain point, not enough to give them full causal status. The truth of the matter is that Aristotle's conception of causality is not the same as ours, hence any embarrassment *we* might feel at Aristotle's remark that places have a certain power is not really germane.

I have been concentrating on the causal role of just two of the dimensions, viz. above and below. If I was right about at least some of the details of right and left, the dimensional places in nature, it is not difficult to see their 'power': the difference they make to the stars is obvious. Dimensional parts also have 'power'. For instance, there was a rather interesting ancient debate about whether the place of fertilization (the right or the left) in a mother's womb determines the sex of the child.[166] These dimensional parts are thus considered to have a potential causal significance, on any account of what is causally significant. Right and left etc. in the universe are not the only naturally occurring distinctions between right and left etc., nor are they the only causally significant ones. Right and left etc. in organisms can also feature in the explanations of certain natural—physical—phenomena.

[164] Algra [56], 151 (Simplicius, *In Phys.* 533. 14–17 = fr. 75. 29–30 Wehrli).
[165] Sorabji [26], 187 n. 6.
[166] See Lloyd [31], with Kember's reply [30] and Lloyd's retort [32].

2

Being In

Introduction: being in a place

In chapter 1 of book IV of the *Physics* Aristotle establishes that there are good reasons for thinking that there are places. His major arguments appeal to replacement (and displacement) and the movement of the natural elements.[1] Both these arguments appeal to *change* of place, which is not surprising, since Aristotle had said from the start that places are important because of change in respect of place, i.e. local motion.[2] To be in local motion is, it seems, to be moving from one place to another—*changing* from having one place to having another. In fact, Plato had already pointed out that rotation is a form of local motion, but does not involve change of place, at least not of the whole body.[3] This issue becomes important when considering the outer sphere of the Aristotelian universe, which is rotating but has no place, and hence cannot be changing place. But it is surely not a bad idea to start off a study of local motion by considering those cases where something goes from having one place to having another, i.e. where something goes from being somewhere to being somewhere else, in the strictly local sense of 'somewhere'.

If something changes place, then it must have a place to change. Something which has a place is (in Greek) *en topōi*, i.e. '*in* a place'. This expression is equivalent to 'somewhere' in the strictly local sense,[4] and is a common Greek idiom. The lack of definite article between *en* and *topōi* (the words for 'in' and for 'place' respectively) might give the impression that the correct translation is 'in place'.[5]

[1] Replacement: 208^b1–8; movement of elements: 208^b8–25.

[2] 208^a31–2.

[3] *Rep.* 436 D 4–E 6.

[4] Cf. *Phys.* III 5, 206^a2–3: ἀλλὰ μὴν τό γε ποὺ ἐν τόπῳ καὶ τὸ ἐν τόπῳ πού.

[5] Ross [3], 366, glossing 205^b31, translates it 'in a place'; thereafter as 'in place'. Hussey [5] translates it 'in a place' at 209^a26, 'in place' everywhere else.

A drawback to this is that the English expression 'in place' means something like 'in its rightful place' (e.g. 'everything is in place: the lookout, the crane, and the getaway car'), but *en topōi* certainly does not mean that—Aristotle cannot possibly think that every located object is in its rightful place, or else he would have trouble expressing the idea that the elements go to their own place. Rather, the best way of translating *en topōi* is 'in a place'. Aristotle says at IV 5, 212^a31–2: 'A body is *en topōi* if, and only if, there is a body outside it which surrounds it.' Then, at IV 5, 212^b8–10, he says: 'The heavens, as has been said, are not, as a whole, somewhere or in a place, since no body surrounds them.' Here he uses the expression *en tini topōi*, which straightforwardly means 'in a place', and it is difficult to see how he could mean something different by *en topōi*.

Being in something in its own right, and being in something derivatively

Those things which are *en topōi* are those things which have places. But things are said to be in (*en*) a place. I take it that Aristotle reasons that whatever *x*'s place is, this much is clear: *x* is *in* it. Aristotle makes the next move in a particularly difficult but interesting passage at the start of *Physics* IV 2:[6]

> Since some things are said in respect of themselves, some in respect of something else, and place may be either (*a*) the common place, in which all bodies are; or (*b*) the proper place which is the first [thing] in which [a body] is (I mean, for example, that you are now in the heavens because you are in the air and that is in the heavens, and you are in the air because you are on the earth and similarly on that because you are in this very place which surrounds nothing more than you), then, if place is the first thing surrounding each body, it will be a kind of limit . . .

In the chapter of which this passage forms the opening lines Aristotle will show that the place of something cannot be identified with either its form or its matter. This passage is supposed to explain how it is that some people have been led from believing (rightly, ac-

[6] 209^a31–$^b2$: *ἐπεὶ δὲ τὸ μὲν καθ' αὑτὸ τὸ δὲ κατ' ἄλλο λέγεται, καὶ τόπος ὁ μὲν κοινός, ἐν ᾧ πρώτῳ (λέγω δὲ οἷον σὺ νῦν ἐν τῷ οὐρανῷ ὅτι ἐν τῷ ἀέρι οὗτος δ' ἐν τῷ οὐρανῷ, καὶ ἐν τῷ ἀέρι δὲ ὅτι ἐν τῇ γῇ, ὁμοίως δὲ καὶ ἐν ταύτῃ ὅτι ἐν τῷδε τῷ τόπῳ, ὃς περιέχει οὐδὲν πλέον ἢ σέ), εἰ δή ἐστιν ὁ τόπος τὸ πρῶτον περιέχον ἕκαστον τῶν σωμάτων, πέρας τι ἂν εἴη* . . . Translation from Hussey [5], 22.

cording to Aristotle) that a place is some kind of limit to believing (wrongly) that the proper place of *x* is *x*'s own limit. By following Aristotle's reasoning, it is possible to discern many important features of his concept of place.

The first thing we need to understand is what the Greek word *en* means. It appears throughout this passage, normally translated 'in', but once translated 'on' (in the expression 'on the earth'). It is clear from the end of the passage that Aristotle identifies '*x* is in *y*' with '*y* contains *x*'. However, there seem to be two types of containing, one which one might call 'circumscriptive', the other 'receptive'.[7] It is 'circumscriptive' containing which interests Aristotle in *Physics* IV 1–5. Circumscriptive containing is when *x* is in its surroundings in the sense that *x*'s surroundings are moulded to *x*; they impinge on *x* and stop at *x*'s borders. In this sense, for instance, liquid is in a vessel; this is indeed Aristotle's favoured illustration of the locative sense of 'in'.[8] In this sense too I am in the air in this room, or a fish is in the sea. One feature of containment of this sort is that when what surrounds *x* is an appropriate medium, movement of *x* will involve displacement of the medium. (A vessel is in fact not such an appropriate medium for its contents, which is what makes it suitable as a vessel.) As I have already emphasized, Aristotle thinks that all locomotion involves displacement, and so it is not surprising that he wishes to concentrate on this kind of containment, i.e. on this way in which one thing is in another. Another feature of this sort of containment is that, for instance, the inner organs of the fish are not in the sea in the same way as the whole of the fish is. The inner organs are remote from the sea.

Receptive containing is quite different. Suppose I am in France; I do not make a me-shaped *hole* in France, nor do I displace parts of France when I travel around in it. When I am in France, my heart and lungs are in France in exactly the same way as any other part of me. As a first approximation, one might say that France is a sort of space which bodies occupy—a sort of interval between given frontiers. Aristotle argues strongly against such a conception of place in *Physics* IV 4, because he thinks that a scientifically adequate account of place could not be constructed along these lines.[9] This is presumably why there is—somewhat surprisingly—no mention

[7] I borrow the terms from Hussey [5], 108.

[8] See *Phys.* IV 3, 210^{a}24. I say much more about the different senses of 'in' below, pp. 67–76.

[9] See Chapter 4 below.

in *Physics* IV 1–5 of *geographical* places. None the less, Aristotle is not wholly uninterested in receptive containing. His examples in the *Categories* of predications in the category of *where* are 'x is in the Lyceum' and 'x is in the agora'.[10] Commentators have argued that Aristotle's conception of place in the *Categories* corresponds more to one of receptive containment—occupying space—than to the more robust conception of the *Physics*.[11] This seems quite true of these examples. The Lyceum and the agora are rather like France—they are spaces or intervals between given frontiers. (If, that is, Aristotle is referring to the *sanctuary* of the Lyceum and not a building.[12]) Although the Lyceum and the agora are not immediately identifiable as *places* as they are defined in the *Physics*, I do not think that this is problematic. The *Categories* is not meant to be a treatise which investigates the definition of (e.g.) places; it investigates the nature of *predication*, and that 'x is in the agora' is a local predication is undeniable. That it should turn out that predications of this sort are not ones that throw light on the nature of places and what it is to be somewhere is of no consequence.

None the less, it is worth exploring a little further the difference between circumscriptive containing and receptive containing. *Circumscriptive* containment comes in two varieties. Take the celestial sphere. It clearly surrounds or encompasses me. In that sense, I am in it (perhaps it would be better to say 'within it', although this is still ambiguous). This is obviously not a case of *receptive* containing, because the celestial sphere is not an interval, parts of which something might occupy. Moreover, it is not fitted to me in the way of normal circumscriptive containment and it is not the case that any movement on my part involves displacing it or parts of it. None the less, such cases of containing are obviously *related* to circumscriptive containing, for if x is in y in this way, there must be a body z (possibly complex) between x and y, such that x is in z (circumscriptively) and z is in y (circumscriptively). So there are two types of circumscriptive containers: those which *contain* proximately (i.e. they are in contact with the contained body, e.g. the air surrounds me proximately), and those which contain *remotely* (e.g. the celestial sphere in so far as it surrounds me remotely). Aristotle

[10] *Cat.* 4, 2^a1–2.

[11] See e.g. Mendell [50]. His argument depends on other considerations than the ones I mention, notably an argument which Aristotle gives in *Cat.* 6 to the effect that places are a continuous quantity.

[12] For details see Lynch [119].

does not always make clear which of the two types of circumscriptive containment he has in mind when he says that something is in something else in the locative sense of 'in'. This actually makes a difference to the very passage under consideration here, as we shall see a little later.

Remote containing and receptive containing can seem similar. Consider the difference between being in France and being in the French Embassy. The Embassy—the building—is a remote container. It is a physical body which defines frontiers such that if anything is within them, it is in that Embassy (even those things which are not in contact with the building itself, e.g. a bird flying around inside). The sense in which x is in the Embassy seems identical to that in which any sublunary body is in (surrounded by) the celestial sphere. The simple thought that anything within the walls of the Embassy is in the Embassy, and the analogous thought that anything within the frontiers of France is in France, might prompt the search for a physical body comparable to the Embassy (the building), in the case of France. But it is no use saying that the countries and oceans which surround France are such a body, since if they were, *they* would have to be identified with France! Furthermore, it is not clear that the frontiers of France are like the walls of the Embassy, since they have no thickness and are therefore not physical bodies. Moreover, France does not stand to its frontiers as the Embassy (a building) stands to its walls, for France stretches *between* its frontiers, whereas the Embassy does not stretch *between* its walls (a building does not stretch between its walls).

It seems, then, that the apparently innocent relation 'x is in y' and its converse 'y contains x' are far from simple. However, there is much that can be said about the opening of *Physics* IV 2 without getting too involved in complex detail; after all, this passage does come at the start of the main body of Aristotle's account of place, and as such should be understandable without bringing in too many of the qualifications which will become necessary later on. In the passage Aristotle distinguishes between the common place *in* which all bodies are,[13] and the proper place of something, which is the first thing *in* which it is.[14] So the suggestion is that any item in (or strictly speaking, *en*) which x is is a place of x. (Later on Aristotle

[13] 209ᵃ32: τόπος ὁ . . . κοινός, ἐν ᾧ ἅπαντα τὰ σώματά ἐστιν.

[14] 209ᵃ33: ὁ δ' ἴδιος, ἐν ᾧ πρώτῳ.

will make a small adjustment to this.[15]) Although Aristotle does not say explicitly in *this* passage that the 'intermediate' items in which I am (between the common place of all things and my proper place) are places, it becomes clear later that this is what he thinks, when he says that a place of a boat on (*en*) a river is the river itself.[16]

The main focus of interest for Aristotle in our passage is the notion of *x*'s *proper* place. The idea behind a proper place is the following. My places differ in that some are more precise than others. The notion of a 'more precise' place is easily defined: *x* is a more precise place than *y* iff anything in *x* is in *y*, but not vice versa. For instance, everything in the air is in the universe, but not vice versa. I am in the universe and in the air, so they are both places of mine, but the air is a more precise place than the universe. Aristotle clearly thinks that I have a *most* precise place, which 'surrounds no more than me' and is 'the first thing in which I am'.[17]

Aristotle's idea, therefore, is to sort through the things in which I am, in order to find the thing in which I am most precisely. He starts with the common place of all things—the broadest, or least precise, of all my places. Then he looks for another item in which I am—another of my places—which is more precise than the common place of all things. (His candidate is the air.) This quest continues until finally the series ends with my *proper* place.

In order to get clearer about this, we must interrupt the argument in order to understand Aristotle's introductory remark that some things are predicated in their own right (*kath' hauto*) and some derivatively (*kat' allo*). Aristotle frequently observes that something holds of something else in respect of itself or in respect of something else;[18] indeed, one of his abiding concerns in his philosophical works is to state precisely what kind of connection there

[15] The adjustment is that the items need to be *immobile*. I discuss this requirement, which comes notoriously late into Aristotle's discussion (IV 4, 212ᵃ18–19), in Chapter 5 below, pp. 155–61.

[16] IV 5, 212ᵃ19–20: *ὁ πᾶς . . . ποταμὸς τόπος*. As Burnyeat remarks ([43], 235 n. 19), 209ᵃ31–209ᵇ1 and a related passage (IV 4, 211ᵃ23–9) 'show Aristotle noticeably happier to use "in" than to use "place" of the broad places intermediate between our proper place and the heavens'. But the river shows that he was content (if not happy) to do so.

[17] 209ᵃ33: *ὁ δ' ἴδιος, ἐν ᾧ πρώτῳ*; 209ᵃ35–ᵇ1: *τῷδε τῷ τόπῳ, ὃς περιέχει οὐδὲν πλέον ἢ σέ*.

[18] The two major texts are *APst.* I 4 and *Meta.* Δ 18. For various reasons, they are not at all easy to reconcile with each other, as Barnes [85], 112, points out, *pace* his fellow Clarendon Aristotelians Kirwan [117], 168–70, and, apparently, Hussey [5], 104.

is between a given object and its properties.[19] But what are predications which hold in respect of the thing itself (or, as I shall sometimes say, 'in its own right'), and predications which hold in respect of something else ('derivatively')?

Broadly speaking, the idea behind 'in respect of itself' and 'in respect of something else' is the following. There are predicates which hold of an item x in virtue of the fact that they hold of some other item, related in some way to x. For instance, Socrates is pale in virtue of the fact that his *surface* is pale, and, one might think, I am knowledgeable in virtue of the fact that *the rational part* of me is knowledgeable.[20] This can be put schematically as follows:

(1) F holds of x in respect of y iff F holds of x in virtue of the fact that F holds of y.

Then we can formulate what it is for F to hold of x in respect of itself:

(2) F holds of x in respect of itself iff F holds of x[21] and there is no further item y such that F holds of x in virtue of the fact that F holds of y.

Note that the 'itself' in 'F holds of x in respect of itself' *could*, grammatically speaking, refer either to F or to x. However, in the sorts of situation which we shall be considering, it refers to x:[22] I am pale in respect of my surface because it is my surface which is pale, and my surface is pale in respect of itself because there is no further item beyond my surface which is pale in virtue of which my surface is pale.

The sorts of relations which hold between x and y such that x is F in respect of y vary according to what F is. In the case of colours, the relation between x and y is that y is the surface of x. To take another example, isosceles triangles have angles which add up to 180° not in respect of themselves, but rather in respect of

[19] The theory of the categories, which on anyone's view of that theory has things to say about this connection, is a particularly elaborate example, and I say more about it in this chapter. See below, 'Having and being in', pp. 76–8.

[20] These two examples come from *Phys.* IV 3, 210^{a}29–30.

[21] You need to repeat the clause 'F holds of x' for the following reason: 'there is no further item y such that F holds of x in virtue of the fact that F holds of y' comes out *true* if F does *not* hold of x.

[22] See Barnes [85], 111 ff., esp. 112, for discussion of cases where the 'itself' in 'F holds of x in respect of itself' may refer to F as well as x. One such case is important in Aristotle's discussion of Zeno's paradox of place; see Chapter 3 below, pp. 89–91.

something else, namely triangles in general.[23] It is not because they are *isosceles* triangles that they have angles which add up to 180°, but because they are *triangles*. Here, the relation between x and y is that of class inclusion, or something like that. (In both these cases the relation could be summed up as 'is part of', and Aristotle does indeed frequently refer to predications in respect of a part.)

Rewriting (1) and (2) as (3) and (4) respectively shows more clearly the role of the relation between x and y:

(3) F holds of x in respect of y iff there is a relation R such that F holds of x because F holds of y and yRx.
(4) F holds of x in respect of itself iff F holds of x and there is no y and no R such that F holds of x because F holds of y and yRx.

In case there is any lack of clarity over the scope of the connectives involved, I shall write (3) and (4) in symbols:

(3_s) $\forall x \forall y \forall F\,[Fx \text{ in respect of } y \leftrightarrow \exists R\,(Fx \text{ because } (Fy \,\&\, yRx))]$.
(4_s) $\forall x \forall F\,[Fx \text{ in respect of } x \leftrightarrow (Fx \,\&\, \neg\exists R\,\exists y\,(Fx \text{ because } (Fy \,\&\, yRx)))]$.

What is the nature of these 'because's? Why is (4) not just (3) but with 'x' written for 'y'? What relations can 'R' stand for? A general study of these sorts of predication would answer these interesting questions, but this is not such a study. The schemata are clear enough to be applied to the predication 'x is in y', which is what Aristotle does in *Physics* IV 2.

Let us therefore resume the argument. Aristotle observes that I am in the common place of all things. Let us for the moment take this to be the universe, i.e. the entire cosmos. Now, the air is in the universe, and so the air is a more precise place than the universe—anything which is in the air is in the universe, but not vice versa. But if I am in the air, then, since the air and the universe are so related, we can infer that I am in the universe. (We cannot, for example, infer that I am in the air from the fact that I am in the universe.) Hence, I am in the universe *because* I am in the air, and the air is a more precise place than the universe. Now, 'x is in y' expresses a relation between x and y. But it also predicates something of x,

[23] I adapt an example of Aristotle's (*APst.* I 4, 73^b32–74^a3 etc.; *Top.* II 3, 110^b22–5). Obviously, my schema oversimplifies in its apparent suggestion that 'F' and 'x' are properties and individuals respectively.

and something of *y*. It predicates *being in y* of *x*, and *being something in which x is* of *y*. (These sorts of property are sometimes called 'relational'.) Take *F* to be the property that 'I am in the universe' predicates of the universe (the property of *being something such that I am in it*). Then, 'I am in the universe because I am in the air and the air is a more precise place than the universe' simply instantiates the right-hand side of our schema (3) above—hence, property *F* holds of the universe *in respect of the air*. In other words, I am in the universe in respect of the air; I am in the universe *derivatively* (*kat' allo*).

When Aristotle says: 'you are now in the heavens because you are in the air and *that is in the heavens*',[24] he means I am in the universe in respect of something else (*kat' allo*), namely the air. And in general, if I am in A and also in B, and A is a more precise place than B, then I am in B derivatively—that is, in virtue of being in A.

The same thing goes for the next step. Anything *en* the earth ('on the earth') is *en* the air ('in the air'), but not vice versa. In this sense, the earth is a more precise place than the air. (This only works if we understand *en tēi gēi* as 'on the earth'.) So I am *en* the air because I am *en* the earth;[25] I am *en* the air derivatively (*kat' allo*). Aristotle does not specify what relation holds between the earth and the air which has the consequence that anything on the earth is in the air, but not vice versa. (In the case of the air and the universe, the air was in fact part of the universe.) It does not much matter how *exactly* the earth and air are related—what counts is that anything *en* the earth is *en* the air, but not vice versa.

Aristotle's next step is (effectively) to claim that there is something, *p*, related to the earth in such a way that anything in *p* is on the earth, but not vice versa. This will mean that I am on the earth derivatively. Moreover, he thinks that there is such a *p* where there is nothing else, *q*, related to *p* in such a way that I am in *q*, and anything in *q* is in *p*, but not vice versa.[26] In other words, Aristotle claims that I have a most precise place, i.e. that there is something

[24] 209^{a}33–4: *σὺ νῦν ἐν τῷ οὐρανῷ ὅτι ἐν τῷ ἀέρι οὗτος δ' ἐν τῷ οὐρανῷ.*

[25] 209^{a}34–5: *καὶ ἐν τῷ ἀέρι δὲ ὅτι ἐν τῇ γῇ.*

[26] Strictly speaking, I should say: 'anything ἐν *q* is ἐν *p*, but not vice versa', But in general, whenever I say 'in' I wish to be taken to translate the Greek word ἐν, whose meaning—as has already been seen, and will be seen again later on—is close to but not the same as that of the English word 'in'.

in which I am non-derivatively (*kath' hauto*).[27] Aristotle refers to it as 'this very place, which surrounds nothing more than you'.

Now, in interpreting Aristotle thus far, I have made some important assumptions. In particular, it is possible to take Aristotle to be saying the following: 'There are the spheres of the fixed stars, of air, and of the earth. Each is in the other. I am in the sphere of earth, which is in the sphere of air, which is in the sphere of the heavens. Thus, I am derivatively in the sphere of the heavens, derivatively in the sphere of the air, and even derivatively in the sphere of the earth, because it turns out that there is something even more precise which contains me which is itself in the sphere of the earth.' The reason why this interpretation is possible is that the word *ouranos* (which appears at 209^{a}33 and 34) is ambiguous between the *whole* of the universe and the sphere of the heavens (the sphere of the fixed stars, the outer sphere of Aristotle's universe).[28] This interpretation also takes *en tēi gēi* to mean 'in the sphere of the earth', which is certainly *possible*. The important difference between this interpretation and mine is that according to this interpretation, Aristotle is 'homing in' on me by specifying ever less remote containers.

There are, however, too many disadvantages to this interpretation. For a start, it is rather strange that Aristotle should miss out the spheres of fire and water. But there are even worse problems. It is clear from the way Aristotle starts the passage that *ouranos* is supposed to pick out the common place of all bodies. However, the celestial sphere is a strange candidate for this common place. For a start, it is a remote container to all sublunary items, and a sort of proximate container to all items in the sphere itself, e.g. the stars. (I say 'sort of proximate container' since the stars are also part of the sphere.) It seems to me undesirable that the common place of all bodies should be related in these two different ways to all bodies. But in fact a remark Aristotle makes in *Physics* IV 5 shows that

[27] Aristotle never talks of a καθ' αὑτό or καθ' ἄλλο place. Nor does he ever *explicitly* say that my proper place is that ἐν which I am non-derivatively. But it is obvious that this is what he has in mind. Note that at 209^{a}33 he says (talking of my proper place) ὁ δ' ἴδιος, ἐν ᾧ πρώτῳ; MS F actually reads πρώτως, which Aristotle often uses to mean καθ' αὑτό. Even if we do not adopt the variant reading, we should none the less understand Aristotle to be explaining ὁ δ' ἴδιος, ἐν ᾧ πρώτῳ in the following lines (cf. λέγω δὲ οἷον, 209^{a}33), in which he shows that the first thing in which *x* is is the most precise thing in which *x* is. But *that* item is that in which *x* is καθ' αὑτό.

[28] See *DC* I 9, 278^{b}10–22; there is a third sense, in which οὐρανός refers to the outermost circumference of the universe (which is also the outermost circumference of the sphere of the heavens).

he takes the universe in the sense of 'everything' to be that thing in which all bodies are,[29] and this strongly suggests that it is the universe which is meant to be the *common* place of everything.

Second, *en tēi gēi* (209^a35) is the Greek for '*on* the earth'. I am *in* the air but *on* the earth. If *en tēi gēi* were to mean 'in the sphere of the earth', it would have to mean *surrounded* by earth, i.e. buried in the earth, but this is not what is meant here. (The sorts of things which are in the earth are the roots of plants, or underground rivers—not the average member of the audience at one of Aristotle's lectures.[30]) Aristotle means that anything which is on the earth is in the air—but not vice versa. We restrict where *in the air* something is when we say that it is *on the earth*, just as we restrict where *in the universe* something is when we say that it is *in the air*.

It might be supposed that the advantage of the second interpretation is that Aristotle is trying to use the transitivity of the relation 'x is in y',[31] i.e. since I am in the sphere of the air and the sphere of the air is in the celestial sphere, I am in the celestial sphere. While it is no doubt true that the relation 'x is in y' (when this relation is understood locatively) is transitive, this does not capture the point of *derivative* containment. For if x is in y in respect of z, it must simply be the case that anything in z is in y but not vice versa—*this may be true even when z is not in y*. Indeed, it turns out that x's proper place, according to Aristotle, is not related to x's surroundings by being in them in the locative way. For instance, something totally surrounded by air is in the air in virtue of its being in that particular inner surface of the air, according to Aristotle. But in that case, the relation between the air and the surface, which has as a consequence that anything within that surface is in the air (but not vice versa), is not the locative 'in', but something like 'is the inner surface of'.[32]

So Aristotle should *not* be interpreted as saying that I am in the outer sphere of the heavens, and in the sphere of the air, and in the sphere of the earth, and then in some particular place in

[29] *Phys.* IV 5, 212^b17: ἐν τῷ οὐρανῷ πάντα· ὁ γὰρ οὐρανὸς τὸ πᾶν ἴσως ('everything is in the universe; for the universe is, of course, the whole').

[30] So I agree with Hussey that Aristotle is not talking of being 'in a subterranean position', and that 'The Greek is idiomatic in the sense "you are on the earth"', but disagree when he says '"the earth" is loosely construed to include the area above the earth's surface' ([5], 104).

[31] A relation R is transitive iff $\forall x\, \forall y\, \forall z\, ((xRy \,\&\, yRz) \rightarrow xRz)$.

[32] *Phys.* IV 4, 211^a26: διὰ τὸ ἔσχατον αὐτοῦ καὶ περιέχον ἐν τῷ ἀέρι φαμὲν εἶναι.

the earth. Rather, he is saying 'I am in the universe, or—more precisely—in the air, or—more precisely—on the earth, or—more precisely—right here, in this place.' Each item introduced restricts where I am. Now, it seems clear from the way Aristotle introduces the passage that each of the items specified which restricts where I am should be considered as one of my places. But although Aristotle uses *en* to express the relation that I have to the universe, the air and the earth, and although this is clearly supposed to be the *en* of *en topōi*, prima facie, I have three quite different relations to the universe, the air, and the earth. I am *en* the air in that I am surrounded more or less completely by it and in contact with it (the air is my proximate container). I am also *en* the universe in that I seem to be 'proximately' surrounded by it, but I am also *included in* it, a part of it. I am *en* the earth; but I am not surrounded by it, even though I am in contact with it. This means that we are still slightly in the dark as to how to go about identifying my place. The fact is that in *Physics* IV 2, 209^a31–209^b2, Aristotle is only at the start of his account of what a place is. It is only to be expected that he will not be totally explicit about what a place is. The crucial point of the passage is to give an informal argument for how it is that we might think that there is such a thing as *x*'s proper place, which surrounds no more than it, and how we might think that this is actually *x*'s limit.

Let us see how Aristotle might take himself to be in a position to show that the item which is the most precise place *en* which *x* is must surround no more than *x*—or even must be a limit.[33] Take my most precise place, *p*. If *p* is not completely fitted to me (i.e. if *p* contains me remotely) then I cannot be in it non-derivatively. For there would be something more precise in which I would be: take, for instance, the walls of this room—they are not fitted exactly to me, and so there is something more precise in which I am, namely the air in the room.[34] Equally, *p*, even if completely fitted to me, must be unextended; suppose it were extended, then there would be something more precise in which I would be, namely some part of *p*. The air in this room affords a rough illustration. It is exactly

[33] 209^a35–$^b1$: τῷδε τῷ τόπῳ, ὃς περιέχει οὐδὲν πλέον ἢ σέ.

[34] Clearly, there are more difficult cases, e.g. when I stand up against the wall. There I am in contact with a combination of wall and air (not to mention the floor, which we have ignored in the example given in the text). I discuss complex cases such as these in Chapter 5 below.

fitted to me, but I am in part of it, e.g. the 'layer' of air 10 cm. thick around me (anything in it is in the air in the room, but not vice versa). But then I am in the layer 5 cm. thick around me, etc.—the most precise part of the air in which I am is the inner *surface* of the air where it is in contact with me.[35]

So it is that if *x*'s proper place is identified with that item which is *x*'s most precise place, i.e. the item in which *x* is non-derivatively, then 'it would be a sort of limit, and this is the reason why people have thought it might be the form or shape of each thing'.[36] When *narrowing down* where *x* is, we try to find something such that it contains nothing more than *x*; so eventually we get to an item which is so precisely fitted to *x* that there is no room for anything else, and which is not even *extended*. And seeing that the surface in question must be precisely fitted to *x*, people have thought that it might actually be the surface or external shape of *x*. Our informal sketch of how this might seem convincing matches the informality of what Aristotle says.

In fact, Aristotle does not think that the proper place of something is its form or surface. I discuss his arguments against this position in Chapter 4 below. But in the opening of *Physics* IV 2, he gets very close to being able to derive some conclusions about what a place is, based entirely on the idea that something is in its proper place *non-derivatively*. He arrives at the notion of a most precise place, a proper place. Anything which is in the common place of all things, namely the universe, has a proper place.[37] All that is missing is some clarity over what 'in' means when we say that something is in its place, and some more clarity over exactly what relation something has to its surroundings.

[35] I am assuming that the surface of something is a part of it.

[36] IV 2, 209^b2: *πέρας τι ἂν εἴη, ὥστε δόξειεν ἂν τὸ εἶδος καὶ ἡ μορφὴ ἑκάστου ὁ τόπος εἶναι.*

[37] It is clearly Aristotle's intention that (1) *x* has at most one proper place (if *x* had two proper places, both must surround *x*, but then one must include the other, in which case one is less precise than the other and is not *x*'s proper place), and that (2) no two things have the same proper place (because something's proper place surrounds no more than it). There could be problems for (2): my proper place is also the proper place of my matter, but I am not the same thing as my matter. Presumably Aristotle would say that matter has a place only in so far as it is related to a composite (form and matter) which has a place.

One thing is said to be in another in many ways

We could be saying one of a large number of things when we say that x is in y. Something is in a place, but we need to know which of the many meanings of 'in' is being used in the expression 'in a place'. If the line of reasoning I have outlined in the first part of this chapter is to be pursued, it is obviously important to be attentive to the nuances of 'in'. A philosophical account which relies heavily on a single concept needs to get clear about the range of meanings of the word which expresses that concept. Indeed, books III and IV of the *Physics* are structured by just such an investigation: Aristotle takes the word *kinēsis*, and notes that its primary meaning is 'motion'; he then supposes that if *kinēsis* also means 'change' quite generally, then this must be by some sort of extension from its primary meaning, and hence that an understanding of what is going on in any change whatsoever presupposes an investigation of the phenomenon of motion.[38]

In fact, the opening of chapter 3 of *Physics* IV, 'Next we must find out in how many ways one thing is said to be *in* another',[39] is typically Aristotelian, for his treatises are full of such investigations of the different aspects of the meaning of some word.[40] *Metaphysics Δ* is nothing but a sequence of chapters laying out the different senses of philosophically important words. In *Metaphysics Γ* 2 Aristotle famously states that 'Something which is is said in many ways' (1003^a33). The *Physics* has its fair share of such passages: *Physics* II 3, 195^a3–4 (=*Metaphysics Δ* 2, 1013^b3–4), 'Therefore causes are said in this number of ways'; *Physics* III 4, 204^a2–3, 'Therefore first we should distinguish in how many ways something *infinite* is said'; and there are many others.

It is impossible to do justice to the difficult question of the doctrine of the *pollachōs legomena* ('things said in many ways') in a few pages. This much seems reasonable: Aristotle isolates the different meanings or uses of a word and tries to show how they are related.[41] The different meanings a word might have do not always arise by chance; and in certain cases they can even be systematically related.

[38] See Chapter 1, 'The importance of motion', for the details of this.

[39] *Phys.* IV 3, 210^a14–15: *μετὰ δὲ ταῦτα ληπτέον ποσαχῶς ἄλλο ἐν ἄλλῳ λέγεται.*

[40] I know of no complete published list of such passages (but see Kirwan [117], 80, for several texts relating to 'being', 'healthy', and 'medical').

[41] See Owen [123] for the introduction of the idea of 'focal meaning'.

The two most common locutions Aristotle uses to express that a word has many meanings are 'something is said to be *F* in many ways', e.g. *Categories* 12: 'one thing is said to be prior to another in four ways'; and '*F* is said in many ways', e.g. *Categories* 15: 'having is said in a number of ways'.[42] The two most common types of locution are the paradigms: for instance, the opening phrase of every chapter of *Metaphysics Δ* is formulated in one of these two ways. (As so often, Aristotle does not distinguish between the use and the mention of a word. So in fact 'one thing is said to be prior to another in four ways' is really a remark about the Greek word *proteron* ('prior'), even though the word is used, not mentioned.)

Aristotle appears to distinguish between the different but unified ways in which something is said and the different but non-unified ways in which something else is said. In *Metaphysics Γ* 2 he says that when investigating the ways in which something is said, we might find that it is said in many ways but 'in relation to one thing and some one nature and not homonymously'.[43]

As an example of the different but unified ways in which something is said, Aristotle takes the predicate 'healthy'. '[E]very healthy thing [is said to be healthy] in relation to health—either from its preserving it, or from its bringing it about, or from being a sign of it, or because it is receptive of it.'[44] Thus, respectively, a diet may be healthy, as may be exercise, red cheeks, and Socrates, but all in different ways. So although 'healthy' has these different meanings, they can all be related one to another. The word is ambiguous—it would, for instance, be perfectly possible to understand what someone means when they say 'Socrates is healthy', but be perplexed when they say 'Walking is healthy'.

This is not a *chance* ambiguity. It can be explained; that is, one can show how the different meanings are related.[45] The different

[42] 14^a26: πρότερον ἕτερον ἑτέρου λέγεται τετραχῶς (14^a26) and τὸ ἔχειν κατὰ πλείονας τρόπους λέγεται (15^b17), respectively. I take it that there is no difference in these contexts between those locutions involving an adverb ending -χῶς and those involving τρόπος. There is a third type of locution: 'F is *n*-fold', e.g. *DA* II 4, 415^b2: τὸ δ' οὗ ἕνεκα διττόν ('that for the sake of which is twofold'), but this is rarer.

[43] 1003^a33–4: πρὸς ἓν καὶ μίαν τινὰ φύσιν καὶ οὐχ ὁμωνύμως.

[44] 1003^a34–7: ἀλλ' ὥσπερ καὶ τὸ ὑγιεινὸν ἅπαν πρὸς ὑγίειαν, τὸ μὲν τῷ φυλάττειν τὸ δὲ τῷ ποιεῖν τὸ δὲ τῷ σημεῖον εἶναι τῆς ὑγιείας τὸ δ' ὅτι δεκτικὸν αὐτῆς.

[45] Sometimes we cannot show this. Aristotle does not reserve ὁμωνύμως for those cases where the meanings cannot be related to each other; sometimes he uses it for any case where something is said in many ways, even if those ways can be unified in some way, e.g. *Cat.* 1, 1^a1–6.

meanings of 'healthy' are related in the following way: to say that something is healthy is none the less to relate it in some way to health. In fact, there seems to be a *primary* sense of 'healthy' (although Aristotle does not say this explicitly), namely that in which things receive or have health (the sense in which an animal or plant is healthy). Let us call items which can receive or have health 'H-items'. Something is healthy in one sense if it keeps H-items healthy; in another sense if it brings it about that H-items are healthy; in another if it is a sign that H-items are healthy. So there is a thesis about language: 'healthy' has different senses, all of which are united around the sense in which H-items are healthy. But there is also an ontological thesis: healthy things (in any of the senses of 'healthy') either just are H-items, or stand in some relevant relation to H-items.

Elsewhere Aristotle says explicitly that there is a primary use of a word, often saying that a particular use is *kuriōtaton* ('most properly so called').[46] What Aristotle is getting at with this word can be shown with the case of *kinēsis* ('motion', 'change'). Aristotle says that the most properly so called of all changes is change of place.[47] Since *all* changes are changes of place *described in a certain way* (but they are not all some other type of change described in a certain way), the different types of 'change' can be shown to connect with change of *place*. The use of *kinēsis* to apply to cases of motion has been *extended* to cover all types of change because of this.

Some of the difficulties involved in this area of Aristotelian doctrine can be seen from the fact that he never actually *says* that *kinēsis* is said in many ways. It appears that Aristotle does not see *kinēsis* as being ambiguous, merely as covering different types of change.[48] None the less, he says that change of place is *kuriōtaton* ('most properly so called'), and it is difficult to make sense of this except against the background of a theory like that of the *pollachōs legomena*. Moreover, in *Physics* II 3 he appears to identify the ways in which something is said to be a cause with types or sorts of cause. The chapter begins: 'Now that these distinctions have been made,

[46] Cf. Frede [100], 178: '[The uses of δύναμις] are related in such a way that there is one basic use, and the others are all derivative from this use and have to be explicated in terms of it (Θ 1, 1046ᵃ9–10).'

[47] *Phys.* IV 1, 208ᵃ31–2. See Chapter 1, pp. 11–15.

[48] Cf. *Cat.* 14, 15ᵃ13–14, where Aristotle lists six types (εἴδη) of κίνησις: (1) changing place; (2) undergoing a qualitative change; (3) growing; (4) diminishing; (5) coming into being; (6) passing away.

we must investigate causes, what they are like and how many there are in number.'[49] Aristotle means how many *types* of causes there are, and this is given by the number of ways in which something is said to be a cause.

When Aristotle discusses the different meanings of *dunamis* in *Metaphysics Θ*, he mentions the *homonymous* ways[50] in which it is said (e.g. its use in geometry)[51] which are derived from some basic use 'by resemblance'.[52] He says: 'We may neglect all the potentialities that are so called homonymously.'[53] However, well into the discussion, Aristotle encourages us to see that a term can apply (non-homonymously) more widely than its primary use, 'by analogy'.[54]

Aristotle seems to envisage that there are cases where a primary meaning has been stretched to give a meaning which is non-homonymously related to the primary one, and cases (even for the same word) where it has been stretched so far that too much of the original primary meaning has been lost for that meaning to be pertinent in a philosophical discussion.[55]

There is some evidence that Aristotle thought of the different meanings a verb can have as being picked out by its compounds. So a verb, V, has various compounds, Prefix + V, which pick out from the large number of possible applications of V some more particular application. In *Metaphysics Δ* 23 he uses compounds of *echein*[56] ('having') to express some of the different ways in which something is said to have something.[57] One way, then, of understanding the ways in which something is said is to think of the

[49] *Phys.* II 3, 194^{b}16–17: διωρισμένων δὲ τούτων ἐπισκεπτέον περὶ τῶν αἰτίων, ποῖά τε καὶ πόσα τὸν ἀριθμόν ἐστιν. See Matthews [121].

[50] *Meta. Θ* 1, 1046^{a}6: ὁμωνύμως.

[51] For this sense see e.g. Plato *Tht.* 147 D–148 B.

[52] *Meta. Θ* 1, 1046^{a}7: ὁμοιότητι.

[53] *Meta. Θ* 1, 1046^{a}6–7: ὅσαι . . . ὁμωνύμως λέγονται δυνάμεις ἀφείσθωσαν.

[54] *Meta. Θ* 6, 1048^{b}7: τῷ ἀνάλογον.

[55] Perhaps Aristotle is getting at something similar when he says (in *Cat.* 15) that the way in which a husband is said to have a wife (and vice versa) is 'most distant' or 'most foreign' (ἀλλοτριώτατος, 15^{b}28) because it just means that they live together, and so does not seem to have anything to do with the principal meaning or meanings of ἔχειν.

[56] οἱ ἀμπεχόμενοι, 1023^{a}11; τὸ περιέχον, τὰ περιεχόμενα, περιέχοντι, 1023^{a}13–14; τὸ συνέχον, συνέχει, 1023^{a}22. See also my remarks about κατέχει at *Phys.* IV 1, 208^{b}4, below, p. 78.

[57] I have more to say about the ways of having, which correspond according to Aristotle to the ways of being in (below, pp. 76–8).

compounds of a verb and the semantic relation between them and the 'core' verb.

Ways of being in

We are now in a position to return to our text:[58]

> Next we must find in how many ways one thing is said to be in another. (1) In one way, as the finger is in the hand, and, generally, the part in the whole. (2) In another, as the whole is in the parts—the whole does not exist apart from the parts. (3) In another, as man is in animal and, generally, species in genus. (4) In another, as the genus is in the species and, generally, the part of the species in the definition. (5) In another, as health is in hot and cold things, and, generally, as the form is in the matter. (6) In another, as the affairs of Greece are in the king [of Persia], and, generally, as things are in the first thing productive of change. (7) In another, as a thing is in its good, and, generally, in its end (that is, the that-for-the-sake-of-which). (8) And—most properly of all so called—as a thing is in a vessel, and, generally, in a place.

Something is said to be *in* another in many ways. This is in the mould of the first locution distinguished above,[59] which brings with it a rather elegant advantage. The two examples I gave for the different locutions ('one thing is said to be *prior* to another in four ways' and '*having* is said in many ways') both concern relations: '*x* is prior to *y*' and '*x* has *y*' respectively. I have written '*x*' and '*y*' in my relations to signal those points where you need to supply a term to make a sentence. Aristotle writes, in the first case, 'one thing' and 'another'. In the second, however, no such help is given: you could follow its wording and say 'walking is said in many ways', 'having is said in many ways', 'giving is said in many ways', and so on, without giving away how many places each of these predicates has (one, two, and three, respectively). If you follow the wording of the first example, you say: 'Something is said to walk in many ways', 'One thing is said to have another in many ways', 'One thing

[58] 210^{a}14–24: *μετὰ δὲ ταῦτα ληπτέον ποσαχῶς ἄλλο ἐν ἄλλῳ λέγεται. ἕνα μὲν δὴ τρόπον ὡς ὁ δάκτυλος ἐν τῇ χειρὶ καὶ ὅλως τὸ μέρος ἐν τῷ ὅλῳ. ἄλλον δὲ ὡς τὸ ὅλον ἐν τοῖς μέρεσιν· οὐ γάρ ἐστι παρὰ τὰ μέρη τὸ ὅλον. ἄλλον δὲ τρόπον ὡς ὁ ἄνθρωπος ἐν ζῴῳ καὶ ὅλως εἶδος ἐν γένει. ἄλλον δὲ ὡς τὸ γένος ἐν τῷ εἴδει καὶ ὅλως τὸ μέρος τοῦ εἴδους ἐν τῷ λόγῳ. ἔτι ὡς ἡ ὑγίεια ἐν θερμοῖς καὶ ψυχροῖς καὶ ὅλως τὸ εἶδος ἐν τῇ ὕλῃ. ἔτι ὡς ἐν βασιλεῖ τὰ τῶν Ἑλλήνων καὶ ὅλως ἐν τῷ πρώτῳ κινητικῷ. ἔτι ὡς ἐν τῷ ἀγαθῷ καὶ ὅλως ἐν τῷ τέλει: τοῦτο δ' ἐστὶ τὸ οὗ ἕνεκα. πάντων δὲ κυριώτατον τὸ ὡς ἐν ἀγγείῳ καὶ ὅλως ἐν τόπῳ.* Translation lightly adapted from Hussey [5].

[59] p. 68.

is said to give another to yet another in many ways', and so on. Thus, saying 'one thing is in another in many ways' has the benefit of showing that being in is a relation between two things.[60]

Now, '*x* is in *y*' is expressed in Greek (as in English) by the verb 'to be' and the preposition 'in'. Thus, there is no simple verb (as there is in the case of 'have') out of which we can make compounds to express the different ways in which something is said to be in something else.[61] However, the syntactic category of the prefix in a verbal compound is, or is very like, an adverb (it takes a verb to make a verb), and it *is* possible to modify the relation '*x* is in *y*' with adverbs or adverbial phrases, and that is exactly what Aristotle does. He says that one *way* (*tropos*)[62] in which something is said to be *in* something else is 'as a finger is in the hand and generally the part in the whole'.[63] Now, the clause introduced by 'as' modifies 'is in' adverbially,[64] and tell us which meanings are in question.[65] The adverbial modifications mark out the different ways in which something may be in something else.

Something is said to be *in* something else in many ways—homonymously, or with reference to one item or primary usage? Well, Aristotle tells us that one of the ways in which *being in* is said, is *kuriōtaton* ('most properly so called'), namely the eighth way, *being in as in a place*.[66] The mere fact that he singles out this use as being the most properly so called is strong evidence that any account of what *unites* all these meanings of 'in' would relate them in some way (not necessarily all in the same way) to this locative meaning.

The meanings of 'in', according to the text of *Physics* IV 3, are

[60] The problem is that you apparently insist that relations hold between different items (that they are irreflexive). Note, however, that Aristotle discusses at length whether something can be in itself (*Phys.* IV 3, 210ᵃ25–ᵇ21), without making any appeal to the fact that he started out with 'one thing is in another in many ways'.

[61] Even if the compound verb ἔνειμι had the same range of meanings as εἰμί ἐν, you could not easily make compounds out of *it*.

[62] τρόπος later came to mean something approximating to an adverb. See Barnes [90].

[63] *Phys.* IV 3, 210ᵃ15–16: ὡς ὁ δάκτυλος ἐν τῇ χειρὶ καὶ ὅλως τὸ μέρος ἐν τῷ ὅλῳ.

[64] It is possible to construe this sentence (and those which follow) as having a λέγεται understood, which ὡς modifies thus: 'in one way, as a finger is said to be in the hand'. But this need not be the case; take *Phys.* IV 3, at 210ᵇ24–6: 'For nothing stops a primary place from being in something, not in it as in a place, but as health is in hot things as a state . . .'. There is no λέγεται here.

[65] Not all adverbial modifications will individuate ways in which something is said to *R* something else. We are interested not in cases of *what the R-ing is like* ('he cut himself badly'), but rather *what the R-ing is* ('he cut off his finger'; cf. 'he cut into his finger', 'he cut up his finger', etc.).

[66] *Phys.* IV 3, 210ᵃ24.

very varied. Each is characterized by an 'as' clause which gives the sense generally (*holōs*); most also have a more precise example of the general sense. I shall list them here, with (in parentheses) the two items which Aristotle gives as examples, in the order *x*, *y*, where *x* is said to be in *y*:

(1) As a part is in the whole. (Example: a finger, the hand.)
(2) As the whole is in its parts. (No example given.)
(3) As the species is in the genus. (Example: man, animal.)
(4) As any part of the species is in its definition. (Example: the genus, the species.)
(5) As form is in the matter. (Example: health, hot and cold things.)
(6) As something is in its primary motive agent. (Example: the affairs of the Greeks, the king of Persia.)
(7) As something is in its end. (Example: something, its good.)
(8) As something is in a place. (Example: something, a vessel.)

I have translated *en* throughout as 'in', although it is evident that some of these locutions cannot be naturally rendered in English in this way. This should not worry us. Simply put, there is no word in English which can be used in all the contexts in which the Greek word *en* can be. On most occasions, however, it can be translated 'in'.

The 'as' clauses giving the general sense of each way in which something is said to be in something else have the form: 'as an F is in a G'. In fact, as a brief glance at the formulations shows, they should be read 'as an F is in *its* G': 'as a part is in *its* whole', 'as the whole is in *its* parts', 'as the species is in *its* genus', 'as any part of the species is in *its* definition', 'as form is in *its* matter', 'as something is in *its* primary motive agent', 'as something is in *its* end', and 'as something is in *its* place (or one of *its* places)'. So we can compile a list of the various relations which hold between two things *x* and *y* such that *x* is said to be in *y*:

(P_1) *x* is in *y* as a part is in its whole iff *x* is a part of *y* and *y* is a whole.
(P_2) *x* is in *y* as a whole is in its parts iff *y* is the set of all the parts of *x*.
(P_3) *x* is in *y* as a species is in its genus iff *x* is a species of the genus *y*.

(P_4) x is in y as a part of the species is in its definition iff there is a species S such that y is the definition G+D ('G' for 'genus', and 'D' for '*differentia*') of S and x is either G or D.
(P_5) x is in y as form is in its matter iff x is the form of the matter y.
(P_6) x is in y as something is in its primary motive agent iff y is the primary motive agent of x.
(P_7) x is in y as something is in its end iff y is the end of x.
(P_8) x is in y as something is in its place iff y is a place of x.

I have written 'iff' in (P_{1-8}), prompted by Aristotle's description of these different ways of being in as 'definitions'.[67] However, there is an apparent objection to this. In *Physics* IV 4, Aristotle says the following:[68] 'So when that which surrounds is not divided from, but rather continuous [with what it surrounds], it is said to be in that thing not as in a place, but as a part in the whole.' Take a certain quantity of water, q. Take a part, p, of this quantity q, such that p is totally surrounded by the rest of q. Aristotle then claims that p is continuous with (and not divided from) that which surrounds it, that is, the quantity of water which corresponds to q without p, i.e. $\{q-p\}$. And in that case, he continues, p is in $\{q-p\}$ *as a part is in the whole*. But p is not part of $\{q-p\}$. This appears to contradict (P_1):

(P_1) x is in y as a part is in its whole iff x is a part of y and y is a whole,

since Aristotle seems to say that p is *in* (in this way) $\{q-p\}$. However, it is probably a mistake to lean too heavily on this isolated text; Aristotle is perhaps just saying (rather clumsily) that p is in q as a part is in the whole when p is continuous with the rest of q.

Aristotle says that the locative sense of 'in' is the primary one. Clearly, there is a close link between this way of being in and the way in which a part is in the whole (the first way), and it is not a large jump from the locative way of being in to the way a property is in an object, or the form is in the matter (i.e. the fifth way)—if Socrates is pale, then pallor resides in Socrates; where Socrates is, there is pallor. And this ties up with the first way as well: pallor is perhaps part of the complex bundle of things which make up

[67] *Phys.* IV 3, 210^{b}9: διορισμῶν.

[68] 211^{a}29–31: ὅταν μὲν οὖν μὴ διῃρημένον ᾖ τὸ περιέχον ἀλλὰ συνεχές, οὐχ ὡς ἐν τόπῳ λέγεται εἶναι ἐν ἐκείνῳ, ἀλλ᾽ ὡς μέρος ἐν ὅλῳ.

Socrates when he is pale. Some of the other meanings of 'in' are special cases of these: the way in which a part is in the whole is connected to the way in which the species is in the genus (the third way); we only have to think of class inclusion, or something like that. And as with non-essential properties of Socrates, so with essential properties: where Socrates (the individual) is, there is his genus, for where Socrates is, there is man. So, where man is, there is animal; the genus is in the species (in the fourth way). (Note that there is a connection here with the first way of being in, since the genus is also in the species as a *conceptual* part.)

This leaves the second, sixth, and seventh ways in which something is said to be *in* something else; they seem to be more remote than the others, at least in this sense: they cannot be idiomatically rendered into English using 'in'. Ross suggests that in English one can say 'the fortunes of a company centre in the managing director' and 'this is the object in which my desires are centred'.[69] As for the second sense, we might say 'the whole consists in its parts'. But it is not clear what the existence of these English paraphrases using 'in' really shows. Two things are relevant here. First, it is clear that the Greek word *en* can express some sort of dependence, so it is idiomatic Greek to say 'such-and-such is in him', meaning that it depends on him, or is up to him.[70] This can be seen in the sixth use of 'in' (i.e. the sense in which something is in its prime mover), and indeed may be the clue to the second sense of 'in', because it is presumably true that a whole *depends* in some way on its parts. Second, and more importantly, the fourth, fifth, sixth, and seventh senses of 'in' mirror the formal, material, efficient, and final causes.[71] Something is *in* its cause in those four ways; this is perhaps connected with the ancient idea that certain causal relations involve the *transmission* of properties.

After a while, one begins to get the feeling that, with a bit of imagination, just about any relation could be shown to be connected to the locative way of being in. It is not clear exactly what the rules are—what counts as an analogous meaning and what does not. However, Aristotle thinks that the locative use of 'in' is that which is appropriate for the relation between things and their places, and therefore that when saying *where* something is, you say in what it is, in the locative sense. But he also thinks that there are items which

[69] Ross [3], 568.

[70] See LSJ s.v. ἐν, A.I.6.

[71] I am grateful to David Sedley for pointing this out to me.

are not in something in the locative sense, *but which are none the less somewhere, because they are in something in some other sense of 'in'*.[72] Aristotle does not think that 'in' is *straightforwardly* ambiguous, any more than 'healthy' is.

Having and being in

In *Metaphysics Δ* 23 Aristotle lists ways in which something is said to *have* something else. At the end of this list he says 'being in something is so called in similar, and corresponding, ways to having'.[73] Aristotle equates '*x* is in *y*' with '*y* has *x*'. The meanings in question often correspond to the nuances of the English 'hold' (often a perfectly respectable translation of *echein*). So a fever is said to *have a hold over* a man, a vessel to *hold* its contents, Atlas to *hold up* the sky, and some things to *hold together* other things—all these italicized expressions translate *echein*. Elsewhere 'have' (or its nuances) gives better sense: someone *has* clothes *on*, matter *has* a form, and a whole *has* parts.[74] The connection between these and some of the ways in which something is said to be in something else is obvious. For instance, the affairs of the Greeks were said to be *en* the king of Persia; so he *echei* ('has a hold over') the affairs of the Greeks.[75] Aristotle groups the ways in which having is said under four headings; it is easy to find in *Physics* IV 3 the corresponding ways in which something is said to be *in* something else, but there seems to be nothing in *Metaphysics Δ* 23 corresponding to (P_2), (P_3), (P_4), and (P_7).[76] Aristotle uses compounds of *echein* throughout the chapter—the one which corresponds to the primary sense of *en* is *periechein*.[77]

[72] This is the subject of the next chapter.

[73] 1023^a23–5: *καὶ τὸ ἔν τινι δὲ εἶναι ὁμοιοτρόπως λέγεται καὶ ἑπομένως τῷ ἔχειν.*

[74] My examples come from 1023^a8–23: fever, $^a10$; vessel, $^a15$; Atlas, $^a20$; holding together (*συνέχει*), $^a22$; clothes, $^a11$; matter, $^a12$–13; whole, $^a17$.

[75] Interestingly, you can do this in English by extending the metaphor slightly: their affairs are *in the hands of* the king; the king *holds* their affairs *in the palm of his hand*.

[76] Aristotle goes through the different nuances of 'part' in *Meta. Δ* 25, and ends up with 'the genus is said to be part of the species, and, in a different way, the species part of the genus' (1023^b24–5). These senses of 'part', coupled with the fact that a whole *has* parts, will yield the converses of (P_3) and (P_4).

[77] Cf. *Meta. Δ* 23, 1023^a14–15: *ἐν ᾧ γάρ ἐστι περιέχοντι, ἔχεσθαι ὑπὸ τούτου λέγεται, οἷον τὸ ὑγρόν φαμεν.* In fact, *περιέχειν* can also be used to cover some of the non-primary senses of *ἐν* (see Solmsen [78], 134).

Of course, Aristotle does not claim to have given an exhaustive list of the ways of having. Indeed, chapter 15 of the *Categories* is another list, this time of eight different ways in which *echein* is said. There are no real surprises given what we have seen so far, although there are further nuances to the familiar meanings of *echein*: for instance, the standard meaning of 'having something on' (e.g. 'I have clothes on') is listed alongside the case where a *part* of me has something on, e.g. when I *have* a ring *on* my finger.[78] It is important to note that Aristotle is uneasy about one of the ways of having, namely that in which a husband has a wife, and a wife has a husband. He says that this amounts to no more than 'lives with', and 'seems to be a very distant way of having'.[79] He is right: analogous meanings are sometimes very far from the original meaning.

In fact, it is the first two ways in which something is said to have something, which are the most interesting. In the first way, something has a state, condition, or quality—for instance, someone *has* knowledge, or *has* virtue.[80] In the second way, something has a quantity—for instance, someone *has* a height of 3 or 4 cubits.[81] The use of the words for 'quality' in connection with the first way and the use of 'quantity' in the second strongly suggest that Aristotle has divided up into two species the way in which *in general* an object has a property. He explains in *Categories* 2 that properties are 'in a subject', when, broadly speaking, the predication in question does not say *what* something is, but rather what it is like, or of what size it is, or to what it is related, or where it is, or when it is, or in what position, or in what state, or what it is doing, or what it is suffering. The way in which properties are in their subjects appears to correspond to the fifth way of being in from *Physics* IV 3;[82] the

[78] *Cat.* 15, 15^b21–3.

[79] 15^b28–9: ἔοικε δὲ ἀλλοτριώτατος ὁ νῦν ῥηθεὶς τρόπος τοῦ ἔχειν εἶναι. I have mentioned this already, n. 55 above.

[80] 15^b17–19: ἢ γὰρ ὡς ἕξιν καὶ διάθεσιν ἢ ἄλλην τινὰ ποιότητα—λεγόμεθα γὰρ ἐπιστήμην ἔχειν καὶ ἀρετήν.

[81] 15^b19–21: ἢ ὡς ποσόν, οἷον ὅ τυγχάνει τις ἔχων μέγεθος—λέγεται γὰρ τρίπηχυ μέγεθος ἔχειν ἢ τετράπηχυ.

[82] See Porphyry, *In Cat.* 78. 6–7: ποῖον οὖν τούτων τῶν ἐννέα λαμβάνεται ἐν τῷ ὑποκειμένῳ ἔν τινι; τὸ ὡς ἐν τῇ ὕλῃ τὸ εἶδος ('Q. To which of these nine senses [of being in] does being in something as a subject belong? A. To the sense in which the form is in the matter'). Translation from Strange [16], 61. Porphyry has *nine* senses because he—intriguingly—divides the eighth sense into 'being in a place' and 'being in a container'. The explanation is that in *Phys.* IV 4 Aristotle distinguishes

converse relation (i.e. the way in which objects *have* properties) was given at *Metaphysics Δ* 23, 'in another way, as something in which something else inheres holds that thing, as a recipient, as bronze has the form of a statue, and a body a disease'.[83]

Aristotle in *Categories* 15 distinguishes the way in which a *quality* is in its subject from the way in which a *quantity* is in its subject. I see no reason why in fact there should not be nine ways of having (and, correspondingly, of being in) corresponding to these two, and the other seven, properties which appear in predications which do not say *what* something is—the so-called *non-substantial* predicates. Things have properties. These properties are of different kinds. Things have these properties in nine different ways, corresponding to the nine types of property. Thus, presumably, properties are in things in nine different ways, although we can unite these nine ways under one 'heading', that of being in as in a subject.

Now, this means that when a place of x is predicated of x in such a way as to say where x is, then that place is in its subject (i.e. x) in the fifth way of being in; thus, objects have (or perhaps 'hold') their places in the way of having which corresponds to the category of *where*. Indeed, at *Physics* IV 1, 208^{b}3–4, Aristotle says that objects hold their place, and his word for 'hold' is *katechei*, which is a compound of *echein* and therefore expresses a way of having[84]—namely, the way in which objects hold, or occupy, their places.

Conclusion: being in a place

What is Aristotle up to? Why does he bother to go through this rigmarole of how things are said to be *in* other things? He does not list all the possible meanings of *en* in *Physics* IV 3—not even all the uses in *Physics* IV, for at several points in the discussion of time in IV 10–14 Aristotle uses the expression 'in time', but there is no hint of a temporal meaning of *en* in IV 3. Presumably he just lists those meanings which he considers pertinent to the enquiry in question.

between x's using something which surrounds it as a place and using it as a vessel (roughly speaking, if y is immobile and surrounds x, then x uses y as a place, if y is moving, then x uses it as a vessel).

[83] *Meta. Δ* 23, 1023^{a}11–13: *ἕνα δ' ἐν ᾧ ἄν τι ὑπάρχῃ ὡς δεκτικῷ, οἷον ὁ χαλκὸς ἔχει τὸ εἶδος τοῦ ἀνδριάντος καὶ τὴν νόσον τὸ σῶμα.*

[84] See above, p. 70.

It is obviously the locative sense of *en* which is in question in the expression *en topōi*, but this is not yet illuminating, for, as we saw before, if *x* is *en y*, then *x* and *y* could be in many different spatial configurations. Rather more illuminating is the apparent identification which we can make between the locative *en* and the verb *periechein*, 'to surround'. For '*y periechei x*' is more restrictive than '*x* is *en y*'. (It seems, for instance, not to be true that the earth *periechei* me, even though I am *en* the earth.) In fact, as I argue in Chapter 5 below, Aristotle takes '*y periechei x*' in the strictest sense of '*y* proximately contains *x*' when he defines what *x*'s place is. However, in *Physics* IV 2–3 no such precision is yet necessary.

Aristotle does *not* simply distinguish the senses of *en* in order to restrict the rest of his discussion to the locative sense, even though it is the primary meaning, and the one of most interest to a study of places. In fact, the whole of IV 3 is dedicated to solving Zeno's paradox of place, for which the distinction between the ways in which one thing is said to be *in* another is essential. The distinction does not just help solve the puzzle; it actually shows how the puzzle is philosophically interesting, and how we *need* a solution which appeals to the different senses of 'in'. But more detailed discussion is needed of what exactly 'surrounding' means—and Aristotle's claim that the most precise (i.e. the proper) place of something is a sort of limit needs a more careful proof than the one sketched above in the first part of this chapter. It is all very well to suppose that something surrounded by air is in the inner surface of the air non-derivatively, but very few things are related to their surroundings so simply. For instance, I am sitting at my desk: I am in contact with the chair, the air, the floor, the keyboard—my immediate surroundings do not seem to have a surface or limit at which they are in contact with me. There needs to be some guarantee that the most precise place in which I am will contain only me.[85]

None the less, the opening of chapter 2 of *Physics* IV can now be elucidated as follows. I am in a number of things (in the locative sense of 'in'). They are my places; it is possible to arrange them in order of how precise they are relative to one another. My most precise place will therefore be such that it is in virtue of being in *that place* that I am in any other place. Let us call any predicate of the form 'is in such-and-such' (where the 'in' is taken in the

[85] I return to these questions in Chapter 5 below.

locative sense) a locative predicate. Then locative predicates hold of x in virtue of the fact that x is in its proper place. But if you predicate a locative predicate of something, then you say where it is, and so that particular fragment of our practice of saying *where* something is depends on the *existence* of proper places. This rather striking and important fact derives from the simple observation that something is *in* a place.

3

Zeno's Paradox of Place

Introduction

IN *Physics* IV 1 Aristotle mentions a puzzle (*aporia*) about places, which he attributes to Zeno of Elea.[1] Aristotle clearly takes this puzzle very seriously, for he devotes an entire chapter (IV 3) to its refutation. Despite this, it is not at all clear what the puzzle actually is, or precisely what its conclusion is supposed to be. Neither Aristotle nor his commentators cite Zeno himself. However, the commentaries of both Philoponus and Simplicius (fifth century AD) contain valuable help, especially where they report the views of Eudemus, a student of Aristotle's, and Alexander of Aphrodisias, the Aristotelian commentator (*fl.* AD *c*.200).

To reconstruct the puzzle which troubled Aristotle is not to reconstruct the argument offered by Zeno. It seems clear from what is known of Zeno's philosophical project that Aristotle has taken the puzzle out of a *longer* Zenonian argument, which in turn formed part of a series of arguments. I am not going to reconstruct Zeno's own argument, for that would require a full study of Zeno's thought: much depends on exactly what form his 'paradoxes' took.[2] To distinguish between the argument which troubles Aristotle and which constitutes a puzzle about places, and the original Zenonian one which contains that first argument as a part and which was probably offered as some sort of paradox, I shall refer to them as 'puzzle' and 'paradox' respectively.

The puzzle needs to be stated; the problem that the puzzle raises needs to be identified; Aristotle's solution needs to be clarified; and his solution tested for adequacy. Some have claimed that Aristotle avoids one infinite regression only to find himself embarking on another—out of the infinite frying pan into the eternal fire.

[1] IV 1, 209ᵃ23–5.

[2] See Barnes [27], chs. XII and XIII *passim*.

The form of the puzzle

Aristotle first mentions the puzzle at *Physics* IV 1, as the fifth of six difficulties[3] which arise for someone investigating places:[4]

(1) Again, if places themselves are among the things which exist, where will they be?[5] For Zeno's puzzle demands some account: for if everything which exists is in a place, it is clear that there will be a place of a place, and this to infinity.

The next and only other time Aristotle mentions the puzzle explicitly is when he gives his solution:[6]

(2) That it is impossible for one thing to be in itself primarily is clear, and the puzzle which Zeno posed, that if places are something, what will they be in,[7] is not difficult to solve: for there is nothing stopping a primary place from being in something else, not, however, as [its occupant] is in that place, but as health is in hot things as a state, and heat is in a body as an affection. Like that, it is not necessary to go to infinity.

As Simplicius pointed out: 'it must be said that in neither of these contexts does he seem to state Zeno's argument in his own words'.[8] The third text which is relevant is where Aristotle restates his solution, although he does not mention Zeno:[9]

[3] Listed at 209^{a}2–29.

[4] 209^{a}23–5 = 29 A 24. 9–11 DK: *ἔτι δὲ καὶ αὐτὸς εἰ ἔστι τι τῶν ὄντων, ποῦ ἔσται; ἡ γὰρ Ζήνωνος ἀπορία ζητεῖ τινὰ λόγον· εἰ γὰρ πᾶν τὸ ὂν ἐν τόπῳ, δῆλον ὅτι καὶ τοῦ τόπου τόπος ἔσται, καὶ τοῦτο εἰς ἄπειρον.*

[5] Ross [3] follows Philoponus, Simplicius, and Themistius ('PST') *πού ἔσται*, which means that the whole sentence runs: 'again, if a place itself is one of the things which exist, it will be somewhere'. But the list of puzzles contains three other questions followed by *γάρ* (209^{a}13–14, 18–19, 26–7), meaning something like 'this is a difficult question, for . . .', and I see no reason not to think that this puzzle was formulated similarly. Cf. Eudemus' comment on Zeno's puzzle, cited below, p. 83.

[6] IV 3, 210^{b}21–7 = 29 A 24. 7–9 DK: *ὅτι μὲν οὖν ἀδύνατον ἐν αὑτῷ τι εἶναι πρώτως, δῆλον· ὃ δὲ Ζήνων ἠπόρει, ὅτι εἰ ὁ τόπος ἐστί τι, ἐν τίνι ἔσται, λύειν οὐ χαλεπόν· οὐδὲν γὰρ κωλύει ἐν ἄλλῳ εἶναι τὸν πρῶτον τόπον, μὴ μέντοι ὡς ἐν τόπῳ ἐκείνῳ, ἀλλ' ὥσπερ ἡ μὲν ὑγίεια ἐν τοῖς θερμοῖς ὡς ἕξις, τὸ δὲ θερμὸν ἐν σώματι ὡς πάθος, ὥστε οὐκ ἀνάγκη εἰς ἄπειρον ἰέναι.*

[7] The manuscripts (and the lemma in Philoponus' commentary) have *ἐν τίνι* (interrogative), whereas Ross [3] prints (with PST) *ἔν τινι*. The problem is parallel to 209^{a}23 (see n. 5).

[8] See Simplicius *In Phys.* 563. 9–10: *ῥητέον ὅτι ἔοικεν ἐν μηδετέρῳ τούτων τῶν χωρίων ἐπ' αὐτῆς θεῖναι τῆς λέξεως τὸν Ζήνωνος λόγον* (translation adapted from Urmson [18], 58).

[9] IV 5, 212^{b}27–9: *καὶ ἔστιν ὁ τόπος καὶ πού, οὐχ ὡς ἐν τόπῳ δέ, ἀλλ' ὡς τὸ πέρας*

(3) And places too are somewhere, not as in a place, but as a limit is in the thing of which it is a limit. For not everything which exists is in a place, but only movable bodies.

Further evidence for what the puzzle is comes from Simplicius, where he cites Eudemus:[10]

But Eudemus gives the following account of Zeno's view: 'Zeno's problem seems to lead to the same point. For it is worth believing[11] that everything that exists should be somewhere; but if places are among the things which exist, where will they be? Presumably in another place, and that in another, and so on.'

The opening premiss of the puzzle is: 'Everything which exists is somewhere'. This much is clear from passages (1) and (3), and Eudemus. Now, the next step cannot be to reason to 'everything which exists is in a place' and then to 'places are in a place'. For Aristotle's solution depends on the fact that things are said to be in one another in different ways. But if the troublesome conclusion is that places are in a place, then adverting to a different sense of 'in' from the locative one will not be much help—it is not that places are *in* a place in some *other* way of being in than the *locative* way of being in. The conscientious Simplicius also spotted this:[12]

if [Zeno] said outright that what is is in a place, how, again, was it possible to solve the problem by the distinction of the ways of being in something, saying (as has just now been said) that a body is in a place but the place is in something else, if it, too, by being one of the things that are, is in a place?

So the opening premiss 'everything which exists is somewhere' does not lead directly to 'everything which exists is in a place'. What, then, *is* the next step? The clue comes from Plato, *Parmenides* 138

ἐν τῷ πεπερασμένῳ. οὐ γὰρ πᾶν τὸ ὂν ἐν τόπῳ, ἀλλὰ τὸ κινητὸν σῶμα. Between 212^b22 and 29 Aristotle gives solutions to four of the six puzzles he had set up in IV 1.

[10] Simplicius *In Phys.* 563. 17–20 = fr. 78. 17–19 Wehrli = 29 A 24. 11–14 DK: *ὁ Εὔδημος δὲ οὕτως ἱστορεῖ τὴν Ζήνωνος δόξαν λέγων· "ἐπὶ ταὐτὸ δὲ καὶ ἡ Ζήνωνος ἀπορία φαίνεται ἄγειν. ἄξιον γὰρ πᾶν τὸ ὂν ποῦ εἶναι· εἰ δὲ ὁ τόπος τῶν ὄντων, ποῦ ἂν εἴη; οὐκοῦν ἐν ἄλλῳ τόπῳ, κἀκεῖνος δὴ ἐν ἄλλῳ, καὶ οὕτως εἰς τὸ πρόσω."*

[11] *ἄξιον.* Barnes [9], 158, reads *ἀξιοῖ* for *ἄξιον*, with Spengel, but the force of Eudemus' remark is that we start from something which merits belief, i.e. an *ἀξίωμα*.

[12] *In Phys.* 563. 4–8: *εἰ γὰρ μὴ ἔν τινι ἀλλ' ἐν τόπῳ ἔλεγε τὸ ὂν ἄντικρυς, πῶς ἔτι δυνατὸν ἦν διὰ τῆς τῶν ἔν τινι διαφορᾶς λῦσαι τὴν ἀπορίαν λέγοντα, ὡς νῦν εἴρηται, ὅτι τὸ μὲν σῶμα ἐν τόπῳ, ὁ δὲ τόπος ἐν ἄλλῳ, εἴπερ καὶ αὐτὸς διὰ τὸ τῶν ὄντων εἶναι ἐν τόπῳ ἐστί.* Translation adapted from Urmson [18], 58.

A 2–B 6. There Parmenides shows (to his interlocutor, the 'young Aristotle' of the dialogue) that if the One is somewhere, then it would be two, hence the One is not somewhere: 'So the one is not somewhere, being neither in itself nor in something else.'[13] In short, 'somewhere' was identified with 'in something'.[14] This needs to be taken in the right way. Plato talks about the One being in contact all around with something which surrounds it.[15] So, in Aristotelian terms, it is the *eighth* way of being in, the local sense of 'in', which is in question.[16] In fact, Plato and everyone else who identified *being somewhere* with *being in something* want the container in question to enclose on all sides and be in contact with the thing contained;[17] in other words, they are interested in what we called 'proximate containing'.

When we look at the argument which Plato gives to Parmenides in the dialogue, we see quite clearly the basis for an infinite regression. If the One is somewhere, then it is in something, hence there will be two: this argument can be extended to show that there will be infinitely many things in one another (given certain assumptions about the relation of being in, e.g. that nothing is in itself). Now, Zeno was a pupil of Parmenides, and part of his project was to defend his thought. Although it is true that we are talking about words put into Parmenides' mouth by Plato, I am inclined to think that this type of argument was current among the Eleatics. There was a traditional identification of being somewhere with being in something (in the local sense); I think that Zeno's puzzle traded on this identification.

That means that the puzzle probably opened as follows:

(1) Everything which exists is somewhere.

But

(2) Everything which is somewhere is in something.

Premiss (1) is already familiar. Premiss (2) is a statement of the identification of being somewhere with being in something. The relation 'x is in y' in (2) is the local 'in'. Although this relation

[13] *Parm.* 138 B 5–6: οὐκ ἄρα ἐστίν που τὸ ἕν, μήτε ἐν αὑτῷ μήτε ἐν ἄλλῳ ἐνόν.

[14] Cf. also Gorgias *On Not Being*, 82 B 3 DK.

[15] *Parm.* 138 A 3–5: ἐν ἄλλῳ μὲν ὂν κύκλῳ που ἂν περιέχοιτο ὑπ' ἐκείνου ἐν ᾧ ἐνείη, καὶ πολλαχοῦ ἂν αὐτοῦ ἅπτοιτο πολλοῖς.

[16] See Chapter 2, p. 76, for the identification of περιέχειν as the converse of the eighth way of being in.

[17] See *Parm.* 138 A 3–7.

seems to have been taken in the *Parmenides* in the sense of 'proximately containing', there is no especial reason to do so here, and it is implausible to imagine that Zeno wished so to restrict the relation '*x* is in *y*'. Both Aristotle (in the second passage cited above) and Eudemus mention a third premiss, 'places exist'.[18] This can obviously be *inferred* from (1) and (2), plus the principle 'whatever something is in exists and is a place'. This principle[19] would then introduce places into the puzzle. But then this is exactly what Aristotle's and Eudemus' premiss does too. So I shall adopt their premiss 'places exist' as step (3).

The third passage of Aristotle cited above suggests that the puzzle contained a step 'places are somewhere'.[20] This is a natural candidate for step (4). Finally, it is concluded that places are in something, as Aristotle says at 210^{b}23 (the puzzle says that 'if places are something, they will be in something'):

(1) Everything which exists is somewhere.

But

(2) Everything which is somewhere is in something.

And

(3) Places exist.

So

(4) Places are somewhere.

So

(5) Places are in something.

For this to give rise to an infinite regression, (5) needs to be taken to mean 'places are in something which exists' (that way, the thing in which a place is will itself—according to premisses (1) and (2)—need to be in something). Hence, (2) should be understood as:

(2) Everything which is somewhere is in something which exists,

and (5) as:

[18] Cf. Aristotle: '*if* places are something . . .' (*Phys*. IV 3, 210^{b}23); Eudemus: '*if* places are among the things which exist . . .' (ap. Simplicius *In Phys*. 563. 19 = fr. 78. 18 Wehrli).

[19] A true principle for Aristotle—recall the opening of IV 2 (discussed in Chapter 2, pp. 55–66).

[20] IV 5, 212^{b}27–9.

(5) Places are in something which exists.

This gives the following overall form to the puzzle:

(1) Everything which exists is somewhere.

But

(2) Everything which is somewhere is in something which exists.

And

(3) Places exist.

So

(4) Places are somewhere.

So

(5) Places are in something which exists.

More machinery is needed to create an infinite regression (not least the premiss that nothing is in itself). But the *general* idea is that the puzzle will give rise to an infinite series of things in one another. Since the 'in' of (2) ('everything which is somewhere is in something which exists') is the local 'in', this means that this infinite series is a series of places of places etc. Hence, Aristotle can say: 'Zeno's puzzle demands some account: for if everything which exists is in a place, then it is clear that there will be places of places.'[21] Aristotle states the puzzle this way for this reason: *if* you interpret the 'in' of (2) as the local 'in', then there will be an infinite series of places.

Aristotle assumes we want to avoid this infinite series (I attempt to explain why below). But in that case, where are places? Everything is somewhere, so *places* have got to be somewhere, but then if 'somewhere' means 'in something' (in the local sense), then places will have places, and so on. So just where exactly *are* places? The puzzle gives rise to this difficulty, as Aristotle (followed closely, as always, by Eudemus) indicates.

The answer—the explanation which Zeno's puzzle needs—is simply that 'somewhere' need not mean 'in something' (in the local sense). As Aristotle says in the third passage, 'somewhere' has a wider meaning (or perhaps better: has other meanings). To

[21] IV 1, 209^{a}24–5: *ἡ γὰρ Ζήνωνος ἀπορία ζητεῖ τινὰ λόγον· εἰ γὰρ πᾶν τὸ ὂν ἐν τόπῳ, δῆλον ὅτι καὶ τοῦ τόπου τόπος ἔσται, καὶ τοῦτο εἰς ἄπειρον.*

be sure, being somewhere is being in something. But since there are different ways in which something is said to be *in* something, there are different ways in which something is said to be *somewhere*.[22] Just as you can be said to be in something *as in a place*, you can be said to be somewhere *as in a place*. Just as you can be said to be in something *as a limit is in the thing of which it is the limit*,[23] so you can be said to be somewhere in this way. Hence, when glossing 'somewhere' as 'in something' (as Zeno and the tradition had done), the 'in' of 'in something' should *not* be understood as covering *solely* the eighth (local) way of being in, on pain of infinite regression.

It is worthwhile remarking that it is philosophically good sense to distinguish between the way in which properties are somewhere from the way in which the objects which they are in are somewhere.[24] I shall be saying more about this below, when discussing Aristotle's solution to the puzzle,[25] but one reason why it is philosophically good sense, according to Aristotle, is that if you do *not* distinguish these different ways of being somewhere, you fall victim to Zeno's infinite series.

So the puzzle allegedly gives rise to an infinite series of things in one another (as Aristotle says in the second passage),[26] and since the 'in' is the local 'in', this is a series of places (as Aristotle says in the first passage).[27] Stating the puzzle in the way I have done accounts for the apparent discrepancy in what Aristotle says, and explains how it is that a *solution* might turn on recognizing that 'somewhere' does not mean 'in a place', but rather 'in something (in any way of being in) in its own right'.

[22] Cf. Eudemus, following Aristotle (ap. Simplicius *In Phys*. 563. 23–4 = fr. 78. 22–3 Wehrli = 29 A 24 DK): *πρὸς δὲ Ζήνωνα φήσομεν πολλαχῶς τὸ ποῦ λέγεσθαι* ('Against Zeno, we shall say that *somewhere* is said in many ways'); Alexander of Aphrodisias also explicitly (*σαφῶς*) stated that 'somewhere' should be taken to have as many meanings as 'in' (Simplicius *In Phys*. 563. 28–33). I take it that Simplicius is keen to report Alexander as having stated this *σαφῶς* because Aristotle does not.

[23] Whichever way of being in this might be.

[24] Cf. Chapter 1, pp. 17–18.

[25] See below, pp. 96–102.

[26] IV 3, 210^b23: *εἰ ὁ τόπος ἐστί τι, ἐν τίνι ἔσται;* If places are in something there will be an infinite series of things in one another; adopt Aristotle's solution and it need not go to infinity: *οὐκ ἀνάγκη εἰς ἄπειρον ἰέναι* (210^b27).

[27] IV 1, 209^a25: *τοῦ τόπου τόπος ἔσται, καὶ τοῦτο εἰς ἄπειρον.*

The problem

Two rather important questions remain unanswered. (*a*) Does the argument really give rise to an infinite series of things in one another—what extra assumptions need to be made? (*b*) If it does, why should this be an *invidious* regress?

(*a*) *Does the argument give rise to an infinite series of things in one another?*

(1) Everything which exists is somewhere.

But

(2) Everything which is somewhere is in something which exists, in the local sense of 'in'.

And

(3) Places exist.

So

(4) Places are somewhere.

So

(5) Places are in something which exists, in the local sense of 'in'.

Of all *Physics* IV 1–5, Zeno's puzzle raises the most difficult interpretative questions—as so often, every time that you think that you have understood one of Zeno's paradoxes, it slips out of your grasp again.

The idea behind the puzzle is surely the following. By (3), places exist, so take some place, p_1; by (1), it is somewhere, and so, by (2), it is in something which exists, p_2. But then, by (1) and (2), p_2 is in something etc. We end up with a series $\{p_1, p_2, p_3, \ldots p_n, \ldots\}$, where each p_i is in p_{i+1}, in the local sense of 'in' (and hence where each p_i is a place). This series 'goes to infinity' (209ᵃ25, 210ᵇ27). What does Aristotle mean by this? He does not just mean that for any p_i, there is something, p_{i+1}, in which it is. He means that there is a *different* thing in which it is; and that for each p_i, a *new* item has to be introduced in which it is.

As it stands, the puzzle has no hope of showing this. It has to be shown that no two consecutive members of the series are identical (i.e. that nothing is in itself), and that there are no *circular* chains of places. Of course, these seem like perfectly plausible assumptions. But this is where things start to get difficult.

The traditional identification of '*x* is somewhere' with '*x* is in something' (in the local sense of 'in') left it open whether *x* could be in itself or something else.[28] Hence, Aristotle observes that *if* nothing is in itself, then each thing is *either* nowhere *or* in something else.[29] This, in conjunction with 'everything is somewhere', entails 'everything is in something else'; Zeno's puzzle must make use of such an assumption in order to generate its infinite series. Hence Aristotle must lead up to his solution of Zeno's puzzle with a discussion of the question of whether something can be in itself or not (210ᵃ25–ᵇ21). That question had already arisen in the *Parmenides*, where Plato had argued both that nothing can be in itself (with the goal of showing that the One is nowhere because it is neither in something else nor in itself),[30] and also that the One is in itself.[31]

Aristotle characteristically argues that in a sense nothing is in itself, but in another sense, something can be in itself. So after observing that if nothing is in itself, then each thing is either nowhere or in something else, he says: 'But this [i.e. being in] is twofold, either non-derivatively or derivatively'.[32] In IV 2 Aristotle had already observed that *x* might be in *y* derivatively or non-derivatively.[33] He argued that *x* is in *y* derivatively (in the local sense) just in case *x* is in *y* but there is something *z* such that *x* is in *z*, and *z* is a more precise place than *y*. However, in IV 3, 210ᵃ25–ᵇ21, Aristotle makes a *different* distinction. In IV 2 '*x* is in *y* in respect of itself' means '*x* is in *y* in respect of *y*' (and '*x* is in *y* in respect of something else' means '*x* is in *y* in respect of something other than *y*').[34] However, in IV 3 '*x* is in *y* in respect of itself' means '*x* is in *y* in respect of *x*' (and '*x* is in *y* in respect of something else' means '*x* is in *y* in respect of something other than *x*').

Thus Aristotle says: 'Whenever the parts of a whole are a container and a thing contained, then the whole is said to be in itself:

[28] See e.g. *Parm.* 138 B 5–6: *οὐκ ἄρα ἐστίν που τὸ ἕν, μήτε ἐν αὑτῷ μήτε ἐν ἄλλῳ ἐνόν.*

[29] *Phys.* IV 3, 210ᵃ25–6: *ἀπορήσειε δ' ἄν τις, ἆρα καὶ αὐτό τι ἐν ἑαυτῷ ἐνδέχεται εἶναι, ἢ οὐδέν, ἀλλὰ πᾶν ἢ οὐδαμοῦ ἢ ἐν ἄλλῳ.*

[30] *Parm.* 138 A 2–B 6. Cf. Owen [75], 92–4.

[31] *Parm.* 145 B 6–C 7.

[32] 210ᵃ26–7: *διχῶς δὲ τοῦτ' ἔστιν, ἤτοι καθ' αὑτό, ἢ καθ' ἕτερον.*

[33] See Chapter 2, pp. 59–63.

[34] See Chapter 2, p. 60.

for ['in'] is said in respect of something's parts too . . .'.[35] The example given for something which is in itself in respect of its parts is a jug of wine. Aristotle's idea is that when you consider the parts of a whole, they might be in one another; but what is true of the part is true of the whole 'in respect of the part', so we can say that the whole is in itself 'in respect of its parts'. Aristotle does not express himself very well,[36] but what is important is that this has nothing to do with derivative containing as seen in IV 2. In IV 2 the discussion centred on the various items of which it is true to say that *x* is in them; in this way, *x* is in different things derivatively (*x*'s broad places), and in one thing non-derivatively (*x*'s proper place). But in IV 3 it is the *parts* of *x* itself and what *they* are in which are under scrutiny—since in some cases the *parts* of *x* are in one another in a certain sense, *x* can be in itself *derivatively*. So there are two types of derivative and non-derivative containing: type (i), which is discussed in IV 2, and according to which nothing can be in itself derivatively or non-derivatively, and type (ii), which is discussed in IV 3, and according to which something *can* be in itself derivatively, but not non-derivatively.

Hence, if Zeno's puzzle is to give rise to an infinite series of places, it cannot use the notion of type-(ii) *derivative* containing. For otherwise, even if it is true that everything which is somewhere is in something which exists, then it is open to someone to say that some things are in themselves, and hence are somewhere. Then, there would be no infinite series of *different* things in one another. Aristotle rightly finds that this would be a rather lame reply to the puzzle, and so he restricts the meaning of '*x* is in *y*' as used in the puzzle to '*x* is in *y* non-derivatively', where this is the second type of non-derivative containing, i.e. we consider the whole of *x* and what it is in, *not* the parts of *x*, and what *they* are in.

So Aristotle concludes that 'it is clear, then, that it is impossible for something to be in itself primarily'[37] (where this applies to *all* the ways in which something is said to be in something).[38] The

[35] IV 3, 210^{a}27–9: *ὅταν μὲν γὰρ ᾖ μόρια τοῦ ὅλου τὸ ἐν ᾧ καὶ τὸ ἐν τούτῳ, λεχθήσεται τὸ ὅλον ἐν αὑτῷ· λέγεται γὰρ καὶ κατὰ τὰ μέρη.*

[36] Hussey's comment on 210^{b}18–21, that it is 'defective and obscure, at best' ([5], 110), applies to much of 210^{a}25–^{b}21.

[37] IV 3, 210^{b}21–2: *ὅτι μὲν οὖν ἀδύνατον ἐν αὑτῷ τι εἶναι πρώτως, δῆλον.*

[38] Cf. 210^{b}8–9: *οὔτε δὴ ἐπακτικῶς σκοποῦσιν οὐδὲν ὁρῶμεν ἐν ἑαυτῷ, κατ' οὐδένα τῶν διορισμῶν* ('When we investigate inductively, we do not see anything in itself, on any of the definitions [of being in]').

'primarily' here (which means—as so often—'non-derivatively'), is designed to rule out cases of type-(ii) derivative containing, according to which it is possible for something to be in itself derivatively.

Let us return to the puzzle:

(1) Everything which exists is somewhere.

But

(2) Everything which is somewhere is in something which exists, in the local sense of 'in'.

And

(3) Places exist.

And

(4) Nothing is in itself.

So

(5) Places are somewhere.

So

(6) Places are in something which exists, in the local sense of 'in'.

I have added premiss (4): nothing is in itself. Does this suffice to give rise to an infinite series of things in one another? There is a series $\{p_1, p_2, p_3, \ldots p_n, \ldots\}$, where each p_i is in p_{i+1}, in the local sense of 'in' (and hence where each p_i is a place). It can be shown (with the help of step (4), 'nothing is in itself') that no p_i is identical with p_{i+1}, but it is not yet possible to show that *in general* no p_i is identical with p_j, where $j \neq i$. However, this is not difficult, since the relation 'x is in y' which is in question here is a transitive relation, and so, since nothing is in itself, no p_i of the series $\{p_1, p_2, p_3, \ldots p_n, \ldots\}$ is identical with p_j, where $j \neq i$.[39]

Zeno's puzzle, in the version which Aristotle chooses to refute, does give rise to a series of things in one another. Hence it gives rise to an infinite series of places of places. Everything which exists is somewhere; everything which is somewhere is in something else which also exists; places exist; so places are somewhere; so every

[39] Cf. Hussey [5], 110: 'To make an infinite regress, Zeno needs at least four assumptions: . . . (c) nothing is in itself; (d) "in" is transitive (to block reciprocal or cyclic containing)'; Barnes [27], 257, premisses (3)–(5).

place is in something else which exists; there can be no 'circular chains' of things in one another; so there is an infinite series of things in one another.

(*b*) *What is unacceptable about the series?*

Aristotle regards this as a *reductio ad infinitum*.[40] The infinite sequence $\{p_1, p_2, p_3, \ldots p_n \ldots\}$ is such that each p_i is in p_{i+1}. On the assumption that the relation 'x is in y' in question is some sort of circumscriptive containing (and we have seen that this is not an unreasonable assumption given the context of the *Parmenides*), what we obtain is effectively a list of ever more remote containers of anything in p_1. In this sense, the puzzle entails that a body has an infinite number of different places (although that may not be a very felicitous way of expressing it—see below). The challenge is to show how this is unacceptable.

Notice that the conclusion of the puzzle does not state that a body has an infinite number of *proper* places, which would otherwise yield a straightforward absurdity: 'how can places be distinct other than by having distinct boundaries? And how can one and the same object . . . have more than one set of boundaries?'[41] The puzzle simply gives us an infinite series of nested items.

But in that case, what *is* the problem with the series? Aristotle never says why we must avoid the series. I want to show how it is that the series would have the disastrous result that it would not make sense any longer to say that something was somewhere, and hence that we could no longer say where anything was, or know where anything was.

To achieve this, I want to return to Aristotle's method of 'homing in' on an object, which we saw in action at the start of *Physics* IV 2. I am in the universe in so far as I am in the air; I am in the air in so far as I am on the earth; I am on the earth in so far as I am *right here*. Each time, I specify more precisely *where* I am. Consider the series in the other direction. I am right here, I am on the earth, I am in the air, I am in the universe. Although Aristotle appears to suggest that each time I am specifying *different* places of mine, this is in fact misleading of him, and may well not correspond to what he actually meant, for it does not seem right to say that each time I am saying that I am somewhere *else*. Rather, I appear to be saying

[40] See Barnes [89], ch. 2, 'Infinite Regressions', for some guidance here.
[41] Barnes [27], 257.

with increasingly less precision where I am. I am specifying where I am—the place in which I am—with increasingly less precision, or more generality. Something which is somewhere has a unique place, not shared by anything else (its proper place), which can be specified in a variety of ways.

This is to be welcomed, since when we ask where something is, we do not always need to know with enormous precision where something is. It is sometimes more useful to know that something is in the kitchen, or in the sea, or whatever, rather than *exactly* where it is. Or again, sometimes we want to know where *two* things are ('Where are the salt and pepper?'—'They are in the kitchen'), and expect as an answer some rather general specification of their respective places which can encompass both. In short, and as noted already, that there are broad and precise places is an important mark of the concept of place.

It should be obvious that Zeno's puzzle puts an end to any hope that where something is could be specified in a *most* general, or *least* precise, way. Why should this matter? Well, Zeno's series gives rise to an infinite predicational chain in the category of *where*. Let us call a series of items $\{P_1, P_2, \ldots P_n, \ldots\}$ a 'predicational chain' in a particular category C iff each P_{i+1}, when predicated of P_i, is a predication in category C. Our series $\{p_1, p_2, p_3, \ldots p_n, \ldots\}$ is such a predicational chain, in the category of where; 'p_i is in p_{i+1}' says where p_i is. But our series is infinite; hence it gives rise to an infinite predicational chain. Aristotle thinks that *in general* there cannot be infinite chains of this sort—that there are no infinite predicational chains in any category. Chapters 19–23 of *Posterior Analytics* I 'form a continuous argument for the thesis that there cannot be predicational chains of infinite length . . .'.[42] Here is a representative passage:[43]

> For items predicated in [the category of] what something is, the case is plain: if it is possible to define anything, or if what it is to be something can be known, and if you cannot survey infinitely many items, then the items predicated in [the category of] what something is must be finite.

Aristotle appears to extend considerations of this sort to all cate-

[42] Barnes [85], 169. See p. 163 for his definition of 'chain', which is not quite the same as the one I gave.

[43] *APst.* I 22, 82^b37–83^a1: *ἐπὶ μὲν οὖν τῶν ἐν τῷ τί ἐστι κατηγορουμένων δῆλον· εἰ γὰρ ἔστιν ὁρίσασθαι ἢ εἰ γνωστὸν τὸ τί ἦν εἶναι, τὰ δ' ἄπειρα μὴ ἔστι διελθεῖν, ἀνάγκη πεπεράνθαι τὰ ἐν τῷ τί ἐστι κατηγορούμενα.* See Barnes [85], 174–5.

gories,[44] but throughout he is ferociously obscure and—probably—fallacious.[45] However, he links the finitude of predicational chains to the possibility of *knowledge*. Aristotle assumes that to know that P is to know why P—to possess the proof that P. Hence he thinks that proofs must be finite, since we are not capable of 'going through' the steps of an infinite proof, and yet we do know things. (Explanations must also be finite; this is familiar from *Metaphysics* α 2.) If they fulfil certain conditions, predicational chains represent the steps in a proof; these chains are called 'demonstrative' predicational chains. Aristotle attempts to show that there are no infinite predicational chains, and hence no infinite demonstrative chains.

One way to see the importance of this is to consider questions again. Suppose I ask *what* Socrates is, and you reply 'a man'. I do not know what a man is, so I ask you, and you reply 'an animal'. But I do not know what an animal is, and so I ask what an animal is, and so on. Each time you introduce a new item in order to help me see what the previous one was. If I am to be able to obtain knowledge in this way, there must be a stop to these questions somewhere along the line. It must be the case that, for answers in such a series of questions to count as *genuine* answers, the series is finite. But since each item in this series says *what* each previous item is, it is a predicational chain in the sense set out above. And hence the thesis that such predicational chains must be finite can be restated as the thesis that such a chain of questions and answers must be finite. Thus Aristotle would insist that such a series of questions be finite.[46]

I want to suggest that Aristotle holds a similar view about where-questions. If I ask where x is, and you give me a reply of the form 'in p_1', then there are surely circumstances in which I do not know where p_1 is, and continue by asking you just that. You reply: 'in p_2'. But I might not know where p_2 is, and so ask where *it* is, etc. If the answers which you give are to count as *genuine* answers, this series of questions and answers must finish somewhere.[47]

[44] *APst.* I 22, 83ᵃ21–3.

[45] Cf. Barnes [85], 181: 'None of the arguments in A 22 is successful; and they cannot be reformulated in such a way as to furnish a proof of Aristotle's contention.'

[46] This means—I suppose—that there is a question (viz. 'but what is a substance?') to which the right reply is not one which introduces a further item.

[47] Questions and answers raise other philosophical difficulties; see Higginbotham [110], with the references there to some modern discussions.

If this is so, then Aristotle must link such a chain of where-questions to the possibility of knowing where x is. One obvious objection to his doing that is that it seems prima facie absurd to claim that in order to know where x is, I have to have knowledge of—as it were—all the links in its locational chain. The same worry arises for all the other categories: 'a child may . . . know what a horse is without knowing what an animal is'.[48] However, in the *Theaetetus* Socrates argues that in order to know what craftsman's clay (a species of clay) is, it is necessary to know what clay itself is,[49] and as Burnyeat points out, this can be made more plausible if the sort of knowledge in question is scientific or (in relevant cases) philosophical knowledge.[50] So perhaps one could make the view I am attributing to Aristotle more plausible by claiming that for adequate scientific knowledge of where e.g. the earth is, it is not enough to know that it is in the air—it is important to know where the air is, and so on. If this is so, then if it is going to be possible to have genuine scientific knowledge of where something is—the sort of knowledge adequate for doing physics—then there must not be an infinite sequence of places in places etc. Put another way, if we are to be able to specify *exactly* where something is, there has to be the possibility of specifying *completely generally* where it is, on pain of not being able to say where it is *at all*.

However, there are more problems than just this with Zeno's infinite sequence. Consider our infinite sequence, $\{p_1, p_2, p_3, \ldots p_n, \ldots\}$. It is an infinite sequence of places. But we have not specified what sorts of places are in question. If any p_i were to be a *proper* place, then the puzzle predicts that it is in something in the local sense of 'in'. But what is a proper place? It is the inner limit of what surrounds something. Suppose *a* is proximately surrounded by a body *b*. The proper place of *a* is the inner limit of *b*. But this inner limit in turn is not in something else in the local sense of 'in'. The inner limit of something is not the right kind of thing which can be in something locally—its relationship to its environment is that of being *continuous* with *b* (or the rest of *b*). Aristotle himself makes the important distinction between (i) when something is not separated but continuous with its surroundings, in which case it is not in them as in a place, and (ii) when something is separated and in contact with its surroundings, in which case it is in them

[48] Barnes [85], 175 (commenting on 82^b37–83^a1). [49] *Tht.* 147 B.

[50] See Burnyeat [94], 387–8.

as in a place.[51] The inner limit of *b* and *b* itself are an example of case (i).

If a *proper* place is to be somewhere, premiss (2) of the puzzle must be false—for proper places cannot themselves have places, neither broad nor proper; if they *are* somewhere, it is not in virtue of having a *place*. This difficulty does not turn on the fact that there is an infinite sequence of places, but arises simply because if proper places exist, then according to the puzzle, they must be somewhere and therefore in something in the local sense of 'in'. However, Aristotle's solution to the infinite regress does actually solve this difficulty as well.

Aristotle's solution

(1) Everything which exists is somewhere.

But

(2) Everything which is somewhere is in something which exists, in the local sense of 'in'.

And

(3) Places exist.

And

(4) Nothing is in itself (nor are there any circular chains of things in one another).

So

[51] *Phys.* IV 4, 211^{a}29–31: *ὅταν μὲν οὖν μὴ διῃρημένον ᾖ τὸ περιέχον ἀλλὰ συνεχές, οὐχ ὡς ἐν τόπῳ λέγεται εἶναι ἐν ἐκείνῳ, ἀλλ' ὡς μέρος ἐν ὅλῳ· ὅταν δὲ διῃρημένον ᾖ καὶ ἁπτόμενον* etc. Unfortunately, Aristotle fudges this distinction in *Phys.* V 3 (and in the parallel passage from *Meta. K* 12). He defines (226^{b}21) things as being coincident (*ἅμα*) iff they are in one primary place (*ἐν ἑνὶ τόπῳ . . . πρώτῳ*); and then two things as being in contact iff their boundaries are coincident (226^{b}23, *ὧν τὰ ἄκρα ἅμα*). It follows that boundaries have primary places, since they can be coincident. Back in *Phys.* IV 4, just a few lines on from the passage cited at the beginning of this note, Aristotle says that the boundaries of two things which are separated and in contact are 'in the same' (*ἐν τῷ αὐτῷ*, 211^{a}33–4). This is clearly meant to be a synonym for *ἅμα*, as Alexander pointed out (ap. Simplicius *In Phys.* 570. 7–9), but presumably Aristotle is trying to avoid using *ἅμα*, in order to avoid the problem of attributing *places* to boundaries. Quite what predicate we are supposed to supply with 'the same' is then a rather interesting question.

(5) Places are somewhere.

So

(6) Places are in something which exists, in the local sense of 'in'.

So there are places in something in something in something *ad infinitum*.

This puzzle creates an infinite series of things in one another, i.e. an infinite sequence of places. Aristotle is worried by the infinity of the places; it is also problematic that places will then have to be somewhere in virtue of being in something (in the local sense of 'in'). This is fine for broad places: I am in the air, and the air is in something in the local sense of 'in'. But primary or proper places cannot be in something in this way, hence according to the terms of the puzzle they are *nowhere*.

But Aristotle thinks that everything is somewhere, and points out that 'there is nothing stopping a primary place from being in something else, not, however, as [its occupant] is in that place, but as health is in hot things as a state, and heat is in a body as an affection'.[52] In other words, primary places *are* somewhere, but not in virtue of being in something in the *local* sense. Aristotle accepts that being somewhere is being in something, and points out that this *holds* for primary places, since they are in something else (for nothing is in itself in any sense of 'in') in some *other* sense of 'in' than the local sense. Premiss (2) is thus false.

Zeno's puzzle identifies 'somewhere' with 'in something' in the local sense of 'in'. Aristotle has shown that 'in' has many senses, and he wants 'somewhere' to have as many senses. In other words, there are many ways in which something is said to be *somewhere*. This seems right. The locative sense of 'in' is the one appropriate for saying where bodies are.[53] But we do not just raise the question 'where?' for bodies. We also raise the question for their parts, and we raise the question for properties. But parts are not related to their surroundings in the way their hosts are, and we can hardly even talk of a property having 'surroundings'. However, other meanings of 'in' (the first way and the fifth way in which something is said to be

[52] *Phys*. IV 3, 210^{b}24–7: *οὐδὲν γὰρ κωλύει ἐν ἄλλῳ εἶναι τὸν πρῶτον τόπον, μὴ μέντοι ὡς ἐν τόπῳ ἐκείνῳ, ἀλλ' ὥσπερ ἡ μὲν ὑγίεια ἐν τοῖς θερμοῖς ὡς ἕξις, τὸ δὲ θερμὸν ἐν σώματι ὡς πάθος.*

[53] Or rather, as Aristotle will qualify later on, the one appropriate for bodies which have something outside of them.

in something else) apply to precisely these cases, and hence the fact that we can talk of parts and properties as being *in* something means that we *can* locate them, in a certain sense. When we say where parts and properties are, we relate them to bodies which themselves have a place. Just as there are some things which are said to be healthy in so far as they promote, maintain, signify health in H-items,[54] so some things are said to be somewhere in so far as they are related to P-items (things which have a *place*).

Parts and properties are not the only items for which the question 'where?' arises. We also ask where sounds or smells are, or where they are coming from;[55] where shadows are; where the battle of Waterloo was, or the cup final; and where Aristotle discusses places other than in the *Physics*. A particularly interesting case is the location of pain—'where exactly is the pain?' is a good preliminary question for a doctor to ask before operating. None of these items is a P-item according to Aristotle. So in order to answer these questions, we have to relate the mentioned items to P-items. Perhaps, therefore, where a sound comes from is where its *source* is; a sound is itself wherever *somebody* can hear it. Hence a first shot at analysing 'I like being somewhere where there is the sound of a stream' might be: 'For all places p, if I am in p and I can hear a stream, then I like being in p', where being in a place is attributed only to P-items (in this case, *me*). Where is wisdom to be found? In Socrates (and hence, more broadly, in Athens). And as for the pain, we want to know in which part of the body the pain is (or rather, which part of the body is in pain, as we misleadingly put it). In short, in linking 'being somewhere' to 'being in something', Aristotle implies that being in a place is the *most proper* way of being somewhere, just as being in a place is the most proper way of being in something.[56] It is then in virtue of the fact that there are things which have a place that we are able to *locate* other sorts of thing, like properties, in *them*.

The consequence of ignoring the different ways in which things are somewhere will be—as Zeno's puzzle shows—to insist that places are somewhere in virtue of having a place; this is nonsense for primary places, as well as for properties etc. Being somewhere is not simply being in something in the local sense of "in", so the

[54] See Chapter 2, p. 69.

[55] O'Shaughnessy [137] has some discussion about the place of sounds.

[56] Cf. IV 3, 210^{a}24.

second premiss of the puzzle is actually *false*. But in that case, Aristotle continues, 'it is not necessary to go to infinity'[57]—without premiss (2), the puzzle is entirely defused. Aristotle's response to the puzzle can therefore be paraphrased as follows: 'Primary places are not somewhere in virtue of being in something in the local sense of "in", hence being somewhere should not be equated with being in something (in the local sense). Hence the puzzle does not arise.'

The puzzle is dead—but reincarnation in a different form is on the cards. One obvious change we could make to the puzzle would be simply to restrict the domain of items under consideration to bodies alone. Obviously, Aristotle holds that all bodies are somewhere, but if 'somewhere' is taken to mean 'is in something else in the local sense', then surely there is an infinite regression. To this Aristotle has the simple answer: not all bodies are somewhere in the local sense. This series of bodies in one another will finally reach the outer sphere of the universe, and that is not somewhere in virtue of being in some other body, as Aristotle often remarks. Hence the premiss 'everything is somewhere', where 'somewhere' is taken strictly in the local sense, will be false, even if we restrict the domain of items under consideration to bodies. The outer sphere of the universe is somewhere in virtue of being in the universe *as a part* (it is in the universe in the first sense of 'in'). And then the *universe* is somewhere, in virtue of being in its parts (in the second sense of 'in').[58] This is why it is so important that 'Aristotle's cosmology can provide a terminal place for all bodies to be permanently in.'[59]

A rather more worrying reincarnation of the puzzle is in the following form:

(1) Everything which exists is somewhere.

[57] 210^{b}27.

[58] Several times in IV 5 Aristotle appears to say that the universe is not somewhere, but he simply means 'somewhere' in the strict local sense. See *Phys.* IV 5, 212^{b}11–12: the universe is in fact somewhere, but κατὰ συμβεβηκός ('accidentally', 'derivatively'), because it is appropriately related to its parts, which *are* in a place. Cf. Taylor [144], 92. Wondering about the spatial location of events, Taylor surmises that 'the sense in which Jones' buttering of the toast can be in the bathroom is distinct from that in which either Jones or the toast can be in it (though doubtless the event derives its ascribed location in some—not entirely straightforward—way from the position of Jones, the toast, or both)', The event is somewhere in virtue of its parts (or perhaps 'components', if you think that events have only temporal parts) being somewhere. Taylor goes on to play the same trick with *sets*. (I am grateful to Kevin Mulligan for drawing my attention to this.)

[59] Burnyeat [43], 232 n. 15. See also Hussey [5], 110.

But

(2) Everything which is somewhere is in something else which exists (in any sense of 'in').

And

(3) Places exist.

So

(4) Places are somewhere.

So

(5) Places are in something which exists (in any sense of 'in').

This is obviously very similar to the original puzzle, but now (2) is given its more 'liberal' interpretation, as Aristotle wishes—'in something' is no longer taken *only* in the local sense of 'in'. If everything is somewhere in virtue of being in something (and nothing is in itself) then surely there will still be an infinite series of items in one another, but just in different senses of 'in'? As Hussey remarks, 'there need be no infinite series of *places*, which may be adequate *ad hominem* against Zeno. But Aristotle still ought to explain how there is not an infinite progression all the same.'[60]

Aristotle will reply as follows: once the second premiss has been 'liberalized', then we need no longer suppose that we are doomed to introduce *new* items which serve to locate items already mentioned in the series. We have already seen how our Zeno series would come eventually to the outer sphere of the universe. This outer sphere is somewhere in virtue of being in the universe as a part. Then the universe is somewhere in virtue of being in its own parts. Here, we are not introducing a *new* item in order to say where the universe is.

The solution to both troubling aspects of the puzzle—the fact that the wrong sorts of items are given a place, and that there is an infinite sequence of places—is to show that being somewhere is not just a question of being in something in the local sense of 'in', but being appropriately related to a body which has a place, by being in it in some sense of 'in'. So, for instance, proper places are in something else—not in the local sense of 'in', but in the sense in which health is in hot things as a state, and in which heat is in a body as an affection. The first of these examples is clearly supposed to

[60] Hussey [5], 110.

invoke the fifth sense of 'in', which Aristotle introduced as follows:[61] '[another way] is as health is in hot and cold things, and in general as the form is in the matter'. But in fact, both of the examples are of the fifth way of being in. For in the *Categories* Aristotle says that a state is one kind of quality[62] (i.e. they are the sort of thing which might be referred to in an answer to the question 'what is *x* like'), and that some affections are qualities,[63] and others belong more properly to the category of undergoing or suffering (i.e. they will feature in answers to the question 'what is happening to *x*?').[64] But any of these properties are in their subject in the fifth way of being in.

Perhaps to our disappointment, Aristotle never actually says *explicitly* what proper places are in. He starts off in IV 3 by suggesting that *x*'s place is in something in the fifth way of being in; but he says explicitly in IV 5 that a place is in what it is in *as a limit is in that of which it is the limit*.[65] From this, and from the official definition of the place of *x* (roughly, the inner limit of the *y* which surrounds *x*),[66] it is obvious that the place of *x* is itself in the *y* which surrounds *x*, since the thing *of which the place is the limit* is the *y* which surrounds *x*. That is, the thing which locates *x*'s place is precisely that item which surrounds *x* whose inner limit *x*'s place is.

The limit of *y* is in *y as a part is in the whole*. (Alexander of Aphrodisias seems to have thought the same.[67]) Aristotle appears to change his mind from saying that places are in something in the fifth way of being in to saying that they are in something in

[61] IV 3, 210ᵃ20–1: *ἔτι ὡς ἡ ὑγίεια ἐν θερμοῖς καὶ ψυχροῖς καὶ ὅλως τὸ εἶδος ἐν τῇ ὕλῃ.*

[62] *Cat.* 8, 8ᵇ26–7: *ἓν μὲν οὖν εἶδος ποιότητος ἕξις.* Presumably also, although this needs careful checking, some states correspond to the category of having (*ἔχειν*').

[63] *Cat.* 8, 9ᵃ28–9: *τρίτον δὲ γένος ποιότητος παθητικαὶ ποιότητες καὶ πάθη.*

[64] At *Cat.* 8, 9ᵇ19–33, Aristotle associates certain types of affection (in particular, fleeting ones) with the category of *πάσχειν*, which is to be expected since *πάθος* is the abstract noun from *πάσχειν*). Later on, however, he appears to reinstate all affections as bona fide members of the category of quality, when discussing change (*Cat.* 14), for there, change of affection seems to count as an alteration (which is explicitly a change of quality).

[65] 212ᵇ27–8: *καὶ ἔστιν ὁ τόπος καὶ πού, οὐχ ὡς ἐν τόπῳ δέ, ἀλλ' ὡς τὸ πέρας ἐν τῷ πεπερασμένῳ.*

[66] I discuss this definition in Chapter 5 below.

[67] Ap. Simplicius *In Phys.* 562. 24–7: *ἀλλὰ καλῶς ὁ Ἀλέξανδρος ζητεῖ κατὰ τίνα τῶν προειρημένων τοῦ ἐν τινι τρόπων ἡ ἐπιφάνειά ἐστιν ἐν τῷ σώματι καὶ ὅλως τὸ πέρας ἐν τῷ περατουμένῳ, πότερον ὡς πάθος ἢ ἕξις ἢ ὅλως ὡς ἐν ὑποκειμένῳ ἢ μᾶλλον ὡς μέρος ἐν ὅλῳ* ('But Alexander rightly raises the question in which of the listed ways of being in something the surface is in the body and in general the boundary in the bounded. Is it as an affection, or condition, or generally being in a substrate, or rather as a part in a whole?') (Urmson [18], 57).

the first way of being in. Alternatively, we could suppose that the examples in IV 3 of health being in hot things and heat being in a body are there just to remind us that 'in' has a wider meaning than the locative 'in'. The idea that a place is in something *as a limit is in that of which it is the limit* connects too closely with the official definition of what a place is—a limit—to be coincidental.

So the proper place of *x* is itself in the *y* which surrounds *x*, as a part is in the whole. I am in the surface of air which surrounds me; but that surface of air is in the air, as a part. It is that relation between the surface and the air which enables me to infer (once we know that I am in that surface of the air) that I am in the air; it is in virtue of the fact that the surface is related to the air in that way that being in the surface of the air—being in my proper place—guarantees that I am in the air. The fact that the air has a place means that it is possible to talk about the whereabouts of my proper place by saying where in the air it is; that is, what its position is in the air. This is a question of the *position* of a part in the whole, and not a question of the *place* of a part.

Aristotle therefore tackles Zeno's puzzle of place by distinguishing different senses of 'in'. Then he shows that nothing is in itself, except for a whole which consists of a container and its contents. That thing is in itself derivatively, in respect of its parts. But that type of derivative containing is not in question in Zeno's puzzle. Rather, the puzzle deals with the standard cases of containing and derivative containing, where, for instance, I am in the air as well as in the inner surface of the air. The puzzle proceeds by stating that since everything is somewhere, and everything which is somewhere is in something in the local sense of 'in' (namely, a place), then places must also be in something, and so on to infinity. The infinite nature of this series appears to render it impossible to say where something is (because infinite chains of genuine where-predications cannot exist). Moreover, it is not true that all places are in a place; proper places are not—furthermore, other items in Aristotle's ontology (properties etc.) are not in a place. But proper places (and properties etc.) are none the less somewhere, hence being somewhere is not a question *solely* of being in a place. Once it is clear what it is to be somewhere, it can be seen that it is *true* that everything is somewhere, but that this does not involve an infinite regression.

4

Three Possibilities for Place: Matter, Form, and Space

Introduction

WE now know a fair amount about Aristotle's concept of place. The investigation of the relation of being in took us far into the analysis of what a place is. Aristotle's official definition of place (which I shall discuss in detail in the next chapter) can be derived from this analysis almost entirely. However, Aristotle's explicit argumentative strategy in *Physics* IV 4 is to put forward four possible identifications of what a place is, to reject three of them, and infer that the fourth must be the correct one. In this chapter I am going to examine why he rejects the three candidates; since the objections brought against the three failed candidates should not stand up against the fourth, it is possible to see some of the implied strengths of Aristotle's official definition of what a place is.

Aristotle says:[1]

> Therefore it is already clear from this what place is. For there are, roughly speaking, four things of which place must be one: either form, or matter, or some extension between the extremes, or the extremes, if there is no extension over and above the magnitude of the body which comes to be [in the place].

He had already rejected in IV 2 the identification of the place of *x* with either the form of *x* or the matter of *x*. In IV 4 he revisits these two identifications (rejecting them again, but giving no new arguments),[2] and rejects the third,[3] that the place of *x* is 'some other

[1] IV 4, 211ᵇ5–9: ἤδη τοίνυν φανερὸν ἐκ τούτων τί ἐστιν ὁ τόπος. σχεδὸν γὰρ τέτταρά ἐστιν ὧν ἀνάγκη τὸν τόπον ἕν τι εἶναι· ἢ γὰρ μορφὴ ἢ ὕλη ἢ διάστημά τι τὸ μεταξὺ τῶν ἐσχάτων, ἢ τὰ ἔσχατα εἰ μὴ ἔστι μηδὲν διάστημα παρὰ τὸ τοῦ ἐγγιγνομένου σώματος μέγεθος.

[2] Form: 211^b10–14; matter: 211^b29–212^a2.

[3] 211^b14–29.

extension always existing over and above the extension of the thing which moves'.[4] And then he simply goes on to infer that place is the fourth:[5]

> If, then, place is none of the three, neither the form nor the matter nor some extension which is always present and different from the extension of the thing that moves, it must be that place is the remaining one of the four, the limit of the surrounding body ⟨at which it is in contact with that which is surrounded⟩.

Since antiquity, commentators have been puzzled by this argument, because Aristotle fails to give an argument for why these four possibilities are exhaustive.[6] However, it is possible to justify the list. Aristotle established early on in IV 1 that places exist—they are different from their occupants and exist in nature.[7] But places are not substances; they are features of substances. Hence, we must find a feature with which to identify them. Either we must identify them with a feature of their occupant, or a feature of what is not their occupant, i.e. the *surroundings* of that occupant. (Take an object; the universe divides into that object and the rest of the universe—i.e. the surroundings of the object.) Given that something's place surrounds or contains it,[8] and is the same size as it,[9] the two candidate features of the object are either the object's matter (which spreads across the object, so to speak; Plato had plumped for this identification in the *Timaeus*) or the object's form (which encloses it). And as far as the object's surroundings go, the only two suitable

[4] IV 4, 212ª3–5: *διάστημά τι ἀεὶ ὑπάρχον ἕτερον παρὰ τὸ τοῦ πράγματος τοῦ μεθισταμένου*. This is the theory which gets close to saying that the place of *x* is the space *x* occupies. I discuss it, and Aristotle's difficult objection, below, pp. 121–32.

[5] 212ª2–6*a*: *εἰ τοίνυν μηδὲν τῶν τριῶν ὁ τόπος ἐστίν, μήτε τὸ εἶδος μήτε ἡ ὕλη μήτε διάστημά τι ἀεὶ ὑπάρχον ἕτερον παρὰ τὸ τοῦ πράγματος τοῦ μεθισταμένου, ἀνάγκη τὸν τόπον εἶναι τὸ λοιπὸν τῶν τεττάρων, τὸ πέρας τοῦ περιέχοντος σώματος ⟨καθ' ὃ συνάπτει τῷ περιεχομένῳ⟩*. The text in angled brackets is Ross's addition (following VPST). See Chapter 5 n. 2 for more.

[6] See Simplicius *In Phys*. 571. 32–3: *ἐὰν μὴ ἀνελλιπὴς ἡ διαίρεσις φανῇ, σαθρὰ δοκεῖ ἡ ἀπόδειξις* ('if the division is not shown to be exhaustive, the proof seems unsound'). Sextus is particularly cunning: he accepts that if places exist, then one of the four definitions must be correct; he accepts Aristotle's rejection of form, matter, and interval, but rejects Aristotle's actual definition, and infers that places do not exist after all (*M* X 29).

[7] 208ᵇ1–25.

[8] This is one of the six basic constraints on an account of place, or *ἀξιώματα*, which Aristotle enumerates at the beginning of *Phys*. IV 4 (210ᵇ34–211ª1).

[9] That something's place (primary place) is the same size as it appears at IV 1, 209ª28–9, as a *supposition* which leads to a puzzle; it then appears as one of the *ἀξιώματα* at IV 4, 211ª1–2.

candidates are the gap in which the object is, or the edges of that gap. Aristotle's four possibilities are thus far from arbitrary.

Form and matter[10]

The form and matter of something might be characterized as follows. Take an object: it has many essential properties, and many non-essential properties. A shiny, hard, small, bronze sphere, for instance, may perhaps also be heavy, moving, at room temperature, etc. Suppose that its hardness is a non-essential property. What then is it exactly which is hard? It cannot be something which includes hardness as one of its components, since (*ex hypothesi*) we could have in front us the same thing, but soft. So there is something which underlies the hardness, something which is subject for the hardness. Which aspects or features of the object are those that underlie its non-essential properties? If that which underlies all the non-essential properties of an object is just that object itself, then this question can be rephrased: what *is* an object?

The answer which Aristotle gives is that it is a certain composite thing, consisting of form and matter,[11] i.e. something which *can* be arranged (disposed, extended), which *is* arranged (disposed, extended) in a certain way. The sort of thing which *can* be arranged is its matter and the *arrangement* is its form, so the bronze sphere consists of matter arranged in a certain form. This composite underlies any changes that something undergoes while retaining its identity. So the bronze sphere may become tarnished over the years, but it still has the same form and matter.

Things sometimes turn into something *else*. A bronze statue might be refashioned into a tureen; some water might turn into air (this was the ancient way of viewing evaporation). These changes—so-called substantial changes—are changes which occur at the level of form or matter. Typically, matter discards one form and takes on another.[12] So if the matter of the bronze sphere becomes arranged

[10] For more details on the metaphysical issues in this section see Bostock [93], Frede [97] and [99], Frede and Patzig [101], and Gill [109].

[11] See e.g. *Phys.* I 7.

[12] Matter is that which underlies (τὸ ὑποκείμενον) a substantial change: for this see *Phys.* I 7, 190^{b}20, 'everything comes to be from that which underlies it and its form' (γίγνεται πᾶν ἔκ τε τοῦ ὑποκειμένου καὶ τῆς μορφῆς). Matter is that which remains (τὸ ὑπομένον): there is no catchy remark which expresses this, but see e.g. *Phys.* I 7,

in a different way (i.e. takes on a different form), the bronze will become something else.

Plato and Aristotle had differing views about these issues. Aristotle thinks that matter is the stuff out of which something is made, and form is the arrangement imposed on this stuff. So the matter of the bronze sphere is, broadly speaking, the bronze out of which it is made, and its form is the form of a sphere: bronze is shaped as a sphere.[13] This composite underlies e.g. the object's change from being untarnished to being tarnished. But if the bronze sphere were refashioned into a cube, this would be a change to its form, and so it will change into something else—namely, a bronze cube. (Suppose the matter of the bronze sphere changes—some alchemist transforms it into a gold sphere—then the bronze sphere changes into something else. None the less, there is some 'deeper' matter, prime matter,[14] which underlies the change from bronze to gold.) Similarly, Aristotle will analyse a house into bricks (matter) and their arrangement (form), imposed by the builder. In the case of a man, things get more complicated. The notions of shape and arrangement might not suffice.

Plato, on the other hand, seems in the *Timaeus* to have thought that the sort of thing which can be arranged is what we might term pure extension, which he identifies with *chōra* ('space'). According to him, the composite of form and matter in the case of the bronze sphere has no bronze-characteristics and is simply a sphere. But a sphere is something spherical. What is it that is spherical? Its extension. (Correspondingly, if the bronze sphere changes into something else, it is the extension which will underlie the change.) The form which bounds this extension will be a shape. (As with Aristotle, if you change the shape, you obtain a new body.) The fundamental things in the world are thus shaped extensions, which then have further properties.[15]

190^{a}24–6: *οὐ μὴν ἀλλὰ καὶ ἐπὶ τῶν ὑπομενόντων ἐνίοτε λέγεται ὡσαύτως· ἐκ γὰρ χαλκοῦ ἀνδριάντα γίγνεσθαί φαμεν, οὐ τὸν χαλκὸν ἀνδριάντα* ('[that something comes to be from something else] is sometimes said in the same way even in the case of things which remain: for we say that a statue comes to be from bronze, and not that the bronze becomes a statue').

[13] The matter is formless, and takes on a form: see *Phys.* I 7, 191^{a}8–12.

[14] There is a huge scholarly debate over this. See Charlton [6], 129–45, with 146–7 of the new edition, for discussion and references.

[15] I explain later in more detail how this sort of theory comes about, and how we can extract it from Plato.

Aristotle focuses in IV 2 on how it is that someone might be tempted to identify the place of *x* with either the form or matter of *x*. According to either the 'Aristotelian' or 'Platonic' way of thinking about an object, the place might appear to be the form (because the place of *x* is a limit, and the form acts as a sort of boundary to the matter). But on the 'Platonic' way of looking at an object, the place may appear to be the extension (which spreads across the object, and is bounded by a certain shape or form).

Place as form or matter

Aristotle provides a sketch of an argument to show that the primary (particular) place of *x* is some sort of limit, the idea being that the primary place of something surrounds nothing more than the object.[16] But if the primary place of *x* is some sort of limit, then which limit is it? Aristotle suggests that someone might opt, not for the (inner) limit of the body containing *x*, but for the limit of *x* itself:[17]

> So place would appear to be the form and the shape of each thing, by which the magnitude and the matter of the magnitude are bounded; for that is what the limit of each thing is. If one considers it in this way, then, place is the form of each thing.

'Magnitude' (*megethos*) is used regularly by Aristotle to denote not only a magnitude in the sense of a quantity, but also something which has a magnitude, i.e. a body. Aristotle is using the word in this second sense here. In a sense the form bounds the object, and in a sense it bounds the matter (much as my skin is *my* boundary, but also the boundary of all that it *encloses*), hence Aristotle says that the form is that 'by which the magnitude *and* the matter of the magnitude are bounded'.[18]

Those that identify the place of an object with its form start from the following idea: the place of *x* is the first thing which encloses *x*. (Even Aristotle agrees with this.) But then they suppose that the first thing which encloses *x* is *x*'s surface. The naturalness of the

[16] See Chapter 2, pp. 65–6.

[17] *Phys.* IV 2, 209^{b}2–6: *ὥστε δόξειεν ἂν τὸ εἶδος καὶ ἡ μορφὴ ἑκάστου ὁ τόπος εἶναι, ᾧ ὁρίζεται τὸ μέγεθος καὶ ἡ ὕλη ἡ τοῦ μεγέθους· τοῦτο γὰρ ἑκάστου πέρας. οὕτω μὲν οὖν σκοποῦσιν ὁ τόπος τὸ ἑκάστου εἶδος ἐστιν.* Translation from Hussey [5].

[18] 209^{b}3–4: *ᾧ ὁρίζεται τὸ μέγεθος καὶ ἡ ὕλη ἡ τοῦ μεγέθους.*

idea that something's surface is its place is reflected in the fact that the surface of something is in geometrical terms a *locus*—in Greek, *topos*.[19] But this surface has no matter and could be identified with the form of the object, for it is just a geometrical shape which bounds the matter of the object.

Having dealt with the identification of *x*'s form with its place, Aristotle then goes on to say that the matter of an object has also seemed to some people a plausible candidate for the place of that object. He explains how this has come about:[20]

> [To those that see] how place seems to be the extension of the magnitude, it is the matter. For [the extension of the magnitude] is different from the magnitude; it is that which is surrounded and bounded by the form, e.g. by a surface or limit, and such is the matter, i.e. the indeterminate: for whenever the limit and properties of a sphere are removed, there remains nothing more than the matter.

Aristotle identifies the urge to equate the place of *x* with the extension (*diastēma*) of *x*. What does *diastēma* mean? Primarily, it means 'interval' or 'distance', e.g. the interval between two notes of music, or two fingers.[21] Thus, it can also mean the interval between the edges of something, and can correspond, according to which edges are in question, to length, breadth, or depth. Hence, Aristotle refers to '[the] three intervals, length, breadth, and depth, by which every body is bounded'.[22] Every body, in other words, has an interval or extension, aspects of which correspond to length, breadth, and depth.[23]

Diastēma may refer to a particular interval (e.g. Euclid I. I: 'let the circle BCD be drawn with centre A and interval [i.e. radius]

[19] See Heath [14], i. 329–31, for *loci* on a surface (τόποι πρὸς ἐπιφανείᾳ); note, however, that not all *loci* are surfaces.

[20] 209ᵇ6–11: ᾗ δὲ δοκεῖ ὁ τόπος εἶναι τὸ διάστημα τοῦ μεγέθους, ἡ ὕλη· τοῦτο γὰρ ἕτερον τοῦ μεγέθους, τοῦτο δ' ἐστὶ τὸ περιεχόμενον ὑπὸ τοῦ εἴδους καὶ ὡρισμένον, οἷον ὑπὸ ἐπιπέδου καὶ πέρατος, ἔστι δὲ τοιοῦτον ἡ ὕλη καὶ τὸ ἀόριστον· ὅταν γὰρ ἀφαιρεθῇ τὸ πέρας καὶ τὰ πάθη τῆς σφαίρας, λείπεται οὐδὲν παρὰ τὴν ὕλην.

[21] See LSJ s.v. διάστημα, Ia, c.

[22] *Phys.* IV I, 209ᵃ4–6: διαστήματα μὲν οὖν ἔχει τρία, μῆκος καὶ πλάτος καὶ βάθος, οἷς ὁρίζεται σῶμα πᾶν.

[23] We are liable to talk as if length, breadth, and depth were adequate for specifying the shape of something. But if I want to describe the shape of Lake Geneva, I am not going to get anywhere with just these three concepts. However, they are adequate for dealing, for example, with many geometrical shapes, and it is probably for this reason that they dominate. Note that Aristotle does not say that they are the *only* extensions.

AB').[24] It may also refer to an *indeterminate* interval: consider the sentence 'the shape of that ball is a sphere' (as opposed to 'that ball is a sphere').[25] If you take the *shape* as the thing which is a sphere (and hence spherical), and the ball as being a sphere derivatively, in virtue of having that shape, then since the shape is a sphere, it must be *something* spherical, namely, spherical interval or extension. On this way of thinking, extension itself is considered as that which is extended as a sphere. But in that case, extension is being considered as something indeterminate, to which form or shape is added to yield something determinate.

Aristotle supposes that somebody might think that for something to be a sphere, then its extension or interval has been extended into the form of a sphere (the size of the sphere would be determined by some further quantity). Thus, just as for something to have length is for it to have extension which is extended lengthwise, so for something to be a sphere is to have extension which is extended as a sphere. This could be reformulated as: a sphere is extension, in the form of a sphere. Here, since extension is what is endowed with a form, form must bound the extension (cf. 209^{b}7–8: the extension 'is what is surrounded and bounded by the form').[26] And hence, extension must be formless and indefinite, a characteristic which it shares with matter (cf. 209^{b}9: 'such is matter, i.e. the indefinite').[27]

The only mystery remaining now is why the extension, even though it shares these properties with matter, should, so conceived, be *identified* as the matter of something. To obtain this picture, we should first of all reflect that if the extension is that which receives the form, Aristotle is going to be keen to think of it as matter. But there is more to this than just a sleight of hand, for Plato, in the *Timaeus*, appeared to claim that objects are definite extensions (i.e. extension + form), endowed with properties.[28] According to

[24] *κέντρῳ τῷ Α, διαστήματι τῷ ΑΒ γεγράφθω κύκλος ὁ ΒΓΔ.* See also LSJ s.v. *διάστημα*, 1b.

[25] In fact, the difference between 'the shape of that ball is a sphere' and 'that ball is a sphere' can also be seen in the differing semantic contribution of 'is'; cf. 'Socrates is 6 foot' and 'Socrates' height is 6 foot', where the first 'is' introduces a quantity belonging to Socrates; the second a species of the genus *height*.

[26] *Phys.* IV 2, 209^{b}7–8: *τὸ περιεχόμενον ὑπὸ τοῦ εἴδους καὶ ὡρισμένον.* Note how this differs from the other Aristotelian conception of extension, where a *body* was bounded by its extension.

[27] *ἔστι δὲ τοιοῦτον ἡ ὕλη καὶ τὸ ἀόριστον.* Cf. *Phys.* I 7, 191^{a}10: the matter of something is *τὸ ἄμορφον*.

[28] See Sorabji [26], chs. 1–7, for some history of this idea.

this view, a ball, for instance, is extension, extended spherically, and then endowed with properties like hardness, bounciness, the colour red, etc. In this analysis, the extension plays exactly the role of matter. This is the view, to which Aristotle implicitly refers, according to which the extension of a magnitude is considered as matter.

This view has come about because if we say that the shape of this ball is spherical, then the shape itself might be thought to be spherical, i.e. extended spherically. So there is extension, extended spherically. But why think that this spherically extended extension is, so to speak, the *substratum* of a body? After all, we can say that the colour of the ball is yellow, and thus think that there is something such as colour, which is coloured yellow. But this does not lead us to think that the *substratum* of the ball is *colour*. The reason why philosophers have thought that extension could be the *substratum* is because there is an asymmetry in the case of extension which there is not in the case of colour or other properties of that sort, namely that while only bodies are coloured, it is not the case that only bodies are extended. For there are geometrical figures, which have length, breadth, and depth, but are not bodies, and hence have no colour etc.[29] On the view we are considering, bodies are assumed to be—roughly speaking—geometrical figures, filled in appropriately. The ball is a sphere (the figure), filled in with properties such as hardness, bounciness, etc.

Let us continue with Aristotle's text. In the passage from *Physics* IV 2 cited above Aristotle says 'whenever the limit and properties of a sphere are removed, there remains nothing more than the matter'.[30] This is in fact a reference to a thought-experiment reported slightly more thoroughly at *Metaphysics Z* 3, 1029ᵃ10–19. The context is the search for substance:[31]

If matter is not a substance, it is hard to see what else could be; for when

[29] Aristotle refers explicitly to the asymmetry in this view, at *Meta. B* 5, 1002ᵃ6–8: *τούτοις γὰρ ὥρισται τὸ σῶμα, καὶ τὰ μὲν ἄνευ σώματος ἐνδέχεσθαι δοκεῖ εἶναι τὸ δὲ σῶμα ἄνευ τούτων ἀδύνατον* ('A body is bounded by these [i.e. surfaces, lines and points]; and they are thought to be capable of existing without body, but a body cannot exist without these'). Oxford translation, [1].

[30] 209ᵇ9–11 (see above, p. 108).

[31] *εἰ γὰρ μὴ αὕτη οὐσία, τίς ἔστιν ἄλλη διαφεύγει· περιαιρουμένων γὰρ τῶν ἄλλων οὐ φαίνεται οὐδὲν ὑπομένον· τὰ μὲν γὰρ ἄλλα τῶν σωμάτων πάθη καὶ ποιήματα καὶ δυνάμεις, τὸ δὲ μῆκος καὶ πλάτος καὶ βάθος ποσότητές τινες ἀλλ' οὐκ οὐσίαι (τὸ γὰρ ποσὸν οὐκ οὐσία), ἀλλὰ μᾶλλον ᾧ ὑπάρχει ταῦτα πρώτῳ, ἐκεῖνό ἐστιν οὐσία. ἀλλὰ μὴν ἀφαιρουμένου μήκους καὶ πλάτους καὶ βάθους οὐδὲν ὁρῶμεν ὑπολειπόμενον, πλὴν εἴ τί ἐστι τὸ ὁριζόμενον*

all else is taken off, nothing apparent remains. For while other things are attributes, products, and capacities of bodies, length, breadth, and depth are quantities and not substances (for a quantity is not a substance). Rather, the substance is that primary thing to which these quantities belong. And yet when length, breadth, and depth are taken away, we see nothing left behind unless there be something which is determined by these, so on this view it must appear that matter alone is substance.

The matter is that which should be left behind (recall that it is that which underlies even substantial changes). This thought-experiment of stripping something of its properties is designed to lead you to that composite which underlies all the properties of an object, and thence to that part which can change (entailing a change in the identity of the object) and which is form, and hence to the other part of the composite, which is matter. According to this passage, when we get to the composite, we see something with length, breadth, and depth—we see something with an extension. Aristotle points out that in fact quantities like length, breadth, and depth *bound* something, and that what they bound is matter. He is neutral as to what the nature of this matter is. What he has said is perfectly consistent with those who hold the sort of view of extension which I have been outlining in this chapter, for they will say that it is extension itself which is given form by length, breadth, and depth—that there is a certain extension, extended lengthwise, breadthwise, and depthwise. Equally, what he says is consistent with his own view, that what is bounded by length, breadth, and depth is the bronze of the sphere.[32]

It does seem that *Metaphysics Z* 3 is aporetic,[33] and so one should be wary of its interpretation. None the less, in *Physics* IV 2 Aristotle

ὑπὸ τούτων, ὥστε τὴν ὕλην ἀνάγκη φαίνεσθαι μόνην οὐσίαν οὕτω σκοπουμένοις. Translation adapted from Bostock [93].

[32] There is an objection to this line of thought (see e.g. Gill [109], 23–6): Aristotle cannot think that bronze could be what is left over at the end of the thought-experiment, because what is left must have no attributes, products, and capacities, whereas matter is for Aristotle something which is typically endowed with, for instance, capacities, notably the capacity to receive a form. But *Meta. Z* 3 talks of stripping away 'attributes, products, and capacities *of bodies*' (1029^{a}13: τῶν σωμάτων πάθη καὶ ποιήματα καὶ δυνάμεις), just as *Phys.* IV 2 talks of stripping away 'attributes of *the sphere*' (209^{b}10: πάθη τῆς σφαίρας), so what we are left with is bronze deprived of those capacities etc. which it shares with the sphere. Hence (perhaps) we can avoid the consequence that what we are left with is not, for instance, hard—and hence is not bronze—by saying that the sphere was not *hard*, but rather *made of something hard*, whereas the matter of the sphere is not *made of something hard*, but is just *hard*.

[33] On this see e.g. Schofield [127].

is clearly referring to the thought-experiment conducted therein, and he does so in order to remind us that it is possible to think that extension would be what remains behind in such an experiment; that is, that extension can be thought to be matter. This is why Aristotle remarks that those who think that the extension of something is its place identify the place of the thing with its matter. (I remind you that this is not the theory which identifies the place of something with some extension which exists independently of *x*—perhaps a segment of space—which is occupied by *x* and is its place. That theory is, to all intents and purposes, the third possibility for what a place is, and is discussed by Aristotle in *Physics* IV 4.)[34]

Aristotle's candidate for the matter of a bronze sphere is the bronze of the sphere, not, in fact, the extension of the sphere. But, nobly, he nowhere makes use of the fact that this is his candidate; his arguments against the identification of matter with place[35] depend entirely on the *logical* properties of what must count as matter, and not on his particular choice of thing which satisfies these properties. Aristotle never says: 'the bronze of the sphere is obviously not its place, hence the identification is wrong'.[36] However, he thinks that others may be tempted to identify the place of something with its extension, if this is taken to be the logical part of a body which is bounded by length, breadth, and depth (i.e. its matter)—this temptation arises for those who view length, breadth, and depth as being the form of something like indeterminate extension.

Aristotle claims that Plato was tempted in this way: that in the *Timaeus* Plato held that the matter of a body is its place;[37] and that he did so because he held a version of the theory that extension underlies bodies as a logical part, and hence as their matter (understood 'neutrally' as in *Metaphysics Z* 3 and *Physics* IV 2); and that he identifies this extension as their place.[38] I am going to claim

[34] See my discussion below, 'Space', pp. 121–32.

[35] For which see below, pp. 119–21.

[36] This is perhaps why some commentators would disagree with me and say that Aristotle in *Phys.* IV 2 (and *Meta. Z* 3) works with a different conception of matter from his normal one. Cf. Gill [109] and Mendell [50], 213 n. 19, who says that Aristotle is not referring to matter in his usual sense of 'whatever underlies elemental change', but rather 'the matter (interval between the limits of the body) and form (limits) of mathematical body (sometimes qua mover)'.

[37] 209^{b}11–12: *διὸ καὶ Πλάτων τὴν ὕλην καὶ τὴν χώραν ταὐτό φησιν εἶναι ἐν τῷ Τιμαίῳ.* Note that *χώρα* is used for 'place'.

[38] 209^{b}12–13: *τὸ γὰρ μεταληπτικὸν καὶ τὴν χώραν ἓν καὶ ταὐτόν.*

that Aristotle is absolutely right to say these things, and that commentators who have jealously kept their Plato uncontaminated by Aristotelian matter have done so at the cost of exegetical propriety.

Plato's *Timaeus*, 48 E–53 C

The *Timaeus* contains some discussion of the status of matter and space. Aristotle hails Plato as the only person to have tried to say what a place is, rather than just affirm that it exists: 'Everyone says that place is something, but what it is, that man alone tried to say.'[39] There is something faintly backhanded about this, for Aristotle will criticize every detail of Plato's account. More significantly, it is not far-fetched to think that Aristotle developed his theory of place precisely in order to take account of a particular inadequacy of Plato's theory, namely its inability to cope with the case of locomotion.[40]

According to the *Timaeus*, the following problem presents itself. The four elements which constitute the world (air, fire, water, and earth) turn frequently into one another, by various physical processes.[41] Thus, for example, fire turns into air, by being extinguished and condensed.[42] Plato remarks, rather obscurely, that it is difficult to be sure that what we call 'air' really is air, and so on, because the elements all turn into one another. What Plato is getting at is that if fire really has become air, then something must have remained to underlie the change, otherwise there would have been no change. When Socrates has been in the sun, white has not become brown: rather, something which is white becomes brown. Similarly, fire does not become air (we were wrong to suppose this):[43] rather, something which was fire is now air. Something fiery became airy. Fire is a *such*, not a *this*.[44]

We thought that fire had become air, but in fact, in order for this to be comprehensible as a change, there had to have been something which persisted through the change.[45] But if fire is a quality (a *such*), what is it a quality of? The answer is: of the 'this', the thing which persists. We should accept the following scheme: there is something

[39] *Phys.* IV 2, 209^b16–17: *λέγουσι μὲν γὰρ πάντες εἶναί τι τὸν τόπον, τί δ' ἐστιν, οὗτος μόνος ἐπεχείρησεν εἰπεῖν.*

[40] See below, p. 119.

[41] 49 B 7–C 7.

[42] 49 C 2–4.

[43] Cf. *ὡς δοκοῦμεν* at 49 B 8.

[44] e.g. 49 D 5–6: *μὴ τοῦτο ἀλλὰ τὸ τοιοῦτον.*

[45] 49 E 2: *ὑπομένον*. Note that Aristotle uses the same word for what remains in a change (*Phys.* I 7, *passim*).

(a this) in which the quality of fire (a such) comes to be.[46] So there is something, in which these qualities come to be: part of it becomes fiery.[47] This thing, or parts of it, is wet, set on fire, and receives the forms of earth and water.[48] The comings and goings of these qualities bring about visible change. What is this thing in which the qualities come to be? It is the receptacle of the qualities, and it is identified at 52 A 8–9 as the space or place in which they come to be. Plato uses the word *chōra* for 'space', probably because it carries connotations of one thing *ceding* to another.[49]

So the theory under consideration suggests that bodies come about because qualities come to be in some section of space or place. This is a version of a theory of place as matter. The place has been identified as that thing which, when endowed with qualities, yields what we call an object (and is thus indissociable from it). In my reconstruction of such a theory, I suggested that the interval should be conceived as receiving form (shape), and then qualities. This is exactly what Plato envisages. In the account of how the four basic elements come about, Plato could not be clearer about the priority of shape:[50]

> In the first place, then, it is of course obvious to anyone that fire, earth, water, and air are bodies; and all body has depth. Depth, moreover, must be bounded by surface; and every surface that is rectilinear is composed of triangles. . . . This we assume as the first beginning of fire and of the other bodies, following the account which combines likelihood and necessity.

So certain stretches of space are marked out, e.g. in the form of triangles; then other properties (perhaps colour, or hardness) come to be in the resultant triangles.

Once, Plato refers to the *chōra* as 'the nature which receives all bodies'.[51] This might appear to be inconsistent with the idea that the receptacle receives *qualities*, since the receptacle plus the quality is supposed to yield a body, and yet here Plato suggests that the receptacle receives a *body*. But the notion of receiving is not as simple

[46] See 49 E 7–50 A 4.

[47] 51 B 4: *πῦρ μὲν ἑκάστοτε αὐτοῦ τὸ πεπυρωμένον μέρος.*

[48] 52 D 5–6: *ὑγραινομένην καὶ πυρουμένην καὶ τὰς γῆς τε καὶ ἀέρος μορφὰς δεχομένην.*

[49] See Chapter 1, p. 23.

[50] 53 C 4–D 6: *πρῶτον μὲν δὴ πῦρ καὶ γῆ καὶ ὕδωρ καὶ ἀὴρ ὅτι σώματά ἐστι, δῆλόν που καὶ παντί· τὸ δὲ βάθος αὖ πᾶσα ἀνάγκη τὴν ἐπίπεδον περιειληφέναι φύσιν· ἡ δὲ ὀρθὴ τῆς ἐπιπέδου βάσεως ἐκ τριγώνων συνέστηκεν. . . . ταύτην δὴ πυρὸς ἀρχὴν καὶ τῶν ἄλλων σωμάτων ὑποτιθέμεθα κατὰ τὸν μετ' ἀνάγκης εἰκότα λόγον πορευόμενοι.* Translation from Cornford [11].

[51] 50 B 6: *τῆς τὰ πάντα δεχομένης σώματα φύσεως.*

as that: I receive a wound, although it is not that five minutes beforehand there is something which is a wound such that I am going to receive it, nor is there something which is going to *become* that wound. If x receives y, it is not necessarily the case that y existed antecedently, or that it makes sense to consider y independently of x.[52]

It is not easy to see what the nature of this *chōra* is, in which properties come to be and perish, and which (in this way) receives objects. It is something essentially shapeless (for it *receives* form), and might therefore seem a rather poor candidate to be a place, since places should be the same shape and size as their occupants. In fact, Plato talks as if there is one huge receptacle (all space, as it were), parts of which receive qualities to yield bodies. Then the whole receptacle should be identified with indeterminate extension, parts of which are singled out by certain shapes which give rise to *placed* geometrical figures, which are in turn filled in by other qualities. One can in this way distinguish between the *chōra tout court*, and then parts of it which underlie particular bodies. These parts already have a form or shape (otherwise it would be impossible to identify them as being the *chōra* occupied by any particular body), and thus, although identifiable as extension, are none the less determinate extension plus quantity.

When x becomes y, the *chōra* which underlies x then underlies y. But it must be one and the same *chōra* which now underlies y and which used to underlie x. Presumably, a stretch of *chōra* is individuated at least by its shape; but the different elements have *differently* shaped pieces of space underlying them. (Plato thinks that earth has a cube underlying it, and fire a pyramid.)[53] How then can we talk of the *same* stretch of *chōra* first being qualified as earth, and then as fire? The simple answer is that one and the same stretch of *chōra* may be marked out at one moment by one form, and at another by two different forms. A cubic section of space can host later on two pyramids, if we imagine the cube split into two along one of its diagonal planes.

[52] Algra [21], 97–8, presents Plato as producing two modes of presentation of the receptacle: one as the thing in which qualities inhere, the other 'as a kind of box in which the things of the phenomenal world are put and through which they move'. He leans heavily on this passage, which I argue does not establish what he thinks it does. The other passages he mentions (60 B–C, 79 D), where Plato talks of bodies moving, are more difficult to explain; none the less, Plato will have to provide some account of locomotion, which surely need not be one which makes the receptacle 'a kind of box' through which bodies move.

[53] See *Tim.* 53 C 4–56 C 7.

Aristotle remarks that Plato identifies the receptacle and *chōra*, and he sees that the *chōra* of Plato's discussion is (i) that which is supposed to persist during elemental change in such a way as to define the change, (ii) that which receives form, and (iii) what underlies an object: in other words, he sees that *chōra* is Plato's candidate for what the matter of something is.[54] As I have already explained, Aristotle's candidate for the matter of something is something like the stuff out of which it is made.[55] Thus their two candidates are very different: Aristotle's happens also to be a candidate for that *out of* which something is made, hence he sometimes calls matter 'that out of which'. But it is not surprising that Plato is not tempted to say that things are actually made out of *chōra* (even if it is a part of the things which occupy it).

Plato's account raises many difficulties. But its most important defect is what it implies about locomotion. I take it that on the view of the *Timaeus*, we cannot say *without qualification* that bodies move around. On that view, a ball, for instance, is an amalgam of spherical space plus other properties, but when it moves and changes place, what happens, presumably, is that all these properties cease to be in that region of space, and then come to be in some neighbouring region of space. Change in respect of place, indeed, no longer holds pride of place, as it does with Aristotle,[56] but is reduced to this *derivative* phenomenon. Indeed, one might go further, and ask what remains in *this* sort of change, such that one could even say that a change has taken place. We would have to appeal to the fact that the same or nearly the same properties are realized in neighbouring regions of space (perhaps with some suitable causal account). In any case, there is no change of place, as it is ordinarily understood. Rather, properties are realized (come to be and perish) in different places at different times. This account may be suitable for the movement of some entities (clouds, for instance), but not for the movement of a ball.

Plato's use of the word *chōra* is interesting. Algra [21], 34, says:

[54] *Phys.* IV 2, 209^{b}11–13: *διὸ καὶ Πλάτων τὴν ὕλην καὶ τὴν χώραν ταὐτό φησιν εἶναι ἐν τῷ Τιμαίῳ· τὸ γὰρ μεταληπτικὸν καὶ τὴν χώραν ἓν καὶ ταὐτόν.*

[55] It is worth pointing out that Plato makes an analogy between the relation of the receptacle to perceived bodies and the relation of gold to the things shaped out of it (*Tim.* 50 A 4–B 5). But it is only an analogy: the receptacle persists through change, just as bronze persists while a sculptor makes different shapes out of it. Also, the receptacle, like the bronze, receives shapes.

[56] See Chapter 1, 'The importance of motion', pp. 11–15.

'we may tentatively conclude . . . that in common parlance *topos* and *chôra* were used more or less *promiscue*, the only traceable difference being that, whereas *chôra* appears to have always denoted a certain *extension*, *topos* could also be used just to denote location in relation to the surroundings.' This is surely right: *chōra* carries the connotation 'that into and out of which things go'.[57] There is an important phrase in the *Timaeus* which points to this slight difference between *topos* and *chōra*:[58] '[We] say that anything that is must needs be somewhere in some place, and occupy some space, and that what is neither on earth nor somewhere in heaven is nothing.' In this passage 'in some place' is quite clearly taken up by 'on earth' (the preposition is the same in each case, *en*), and 'occupy some space' is taken up by 'somewhere in heaven' (the preposition *kata* is used both times, appearing the first time as the prefix in *katechon*).[59] In other words, there is precisely the difference noted between *topos* and *chōra*, illustrated by the fact that English uses different prepositions: 'on earth' and 'in heaven'. *Topos* is used for 'solid region', or 'land mass', and *chōra* (roughly speaking) for 'air' or 'space'. In ordinary Greek, therefore, the two words have very slightly different meanings, and it is worth preserving the slightly artificial renderings 'place' and 'space' in order to show this. Of course, Plato does not offer two different 'analyses' for '*x* is in a place' and '*x* occupies some space', nor does Aristotle. The vast majority of the time, Plato uses *chōra* for the receptacle, but he does not flinch from saying (at 52 A 6, for instance) that the qualities which come to be in the receptacle come to be in some *place*, using the word *topos*. Inversely, Aristotle uses *topos* almost exclusively, but occasionally uses *chōra*, especially towards the beginning of his discussion (e.g. IV 1, 208^b7, 32, and 209^a8). Presumably, as his theory takes shape, he uses *topos* precisely in order to bring out the difference between his account and Plato's.[60]

In IV 2 Aristotle uses *chōra* to pick out Plato's candidate for what a place is, and *topos* for his own; hence he can say that even in the unwritten doctrines '[Plato] declared that *topos* and *chōra* were

[57] See Chapter 1, p. 23.

[58] 52 B 3–5: *πρὸς ὃ δὴ καὶ ὀνειροπολοῦμεν βλέποντες καί φαμεν ἀναγκαῖον εἶναί που τὸ ὂν ἅπαν ἔν τινι τόπῳ καὶ κατέχον χώραν τινά, τὸ δὲ μήτ' ἐν γῇ μήτε που κατ' οὐρανὸν εἶναι.*

[59] Note that at *Phys.* IV 1, 208^b3–4, Aristotle says that when something replaces something else then it *κατέχει . . . τὸν αὐτὸν τόπον*. See also Chapter 2, p. 78.

[60] I am grateful to Francis Wolff for pressing this home to me.

the same thing'.[61] In the unwritten doctrines Plato still proposed a theory of place as receptacle. (As Aristotle clarifies in a parenthesis later on in IV 2, in the unwritten doctrines, the receptacle was elucidated in a slightly different way from that of the *Timaeus*.)[62] He takes himself to be offering a *replacement* for Plato's theory, as can be seen from the fact that he says of Plato: 'Everyone says that place [*topos*] is something, but what it is this man alone has tried to say.'[63] Aristotle sees Plato's theory, despite its differences from his, as a theory of place or *topos*.

Aristotle is right that Plato, in the *Timaeus*, offers a theory of place which identifies place with extension, taken in such a way that it is the matter of something. However, it is possible to criticize Aristotle's interpretation of Plato on two points, at least.

First, Aristotle claims that Plato said that the Forms and numbers were not in a place, but Aristotle claims that since they come to be in the receptacle, they are thereby in a place, and hence Plato was wrong to insist that Forms and numbers were not in a place.[64] However, in the *Timaeus* the shapes and qualities which come to be in the receptacle are *copies*[65] of the eternal Forms. Plato has to say this. For these qualities come to be and pass away in the *chōra*, and so cannot be eternal; moreover, if it were the Form itself of, say, hardness which came to be somewhere, to yield a hard body, then there could not be some other hard body somewhere else at the same time, for there is only one Form of hardness.[66] So I do not think that Aristotle is right to say that according to the *Timaeus* the Forms are in a place because they come to be in the receptacle: the fact that the *copies* come to be in the receptacle is not evidence for that view. (There may of course be evidence in *other* dialogues for the view that the Forms are somewhere.) This small misunderstanding is of no importance to Aristotle's rejection of Plato's account.

Second—more importantly—Aristotle takes no account of the fact that Plato's description of the receptacle and its relation to fire, earth, etc. is supposed to be a description of a state of the universe

[61] *Phys.* IV 2, 209^b15–16: *ὅμως τὸν τόπον καὶ την χώραν τὸ αὐτὸ ἀπεφήνατο.*

[62] IV 2, 209^b35–210^a2: *τὸ μεθεκτικὸν ὁ τόπος, εἴτε τοῦ μεγάλου καὶ τοῦ μικροῦ ὄντος τοῦ μεθεκτικοῦ εἴτε τῆς ὕλης, ὥσπερ ἐν τῷ Τιμαίῳ γέγραφεν.*

[63] 209^b16–17.

[64] 209^b33–5.

[65] 51 A 2: *ἀφομοιώματα.*

[66] Compare the fact that for Aristotle bodies and properties are both somewhere but in different ways, such that a body cannot be both somewhere and somewhere *else* simultaneously, whereas properties can.

before the Demiurge comes along and shakes it up and orders it. Quite what hangs on this is unclear, but it does mean that it may be unfair to Plato to press certain details of his account as if it were intended to be definitive.

Why is place not form or matter?

Aristotle's central argument against the identification of the place of something with its form or matter runs as follows:[67]

> The form and the matter are not separated from the object, but the place may be: in that in which air was, water comes to be, as we said, when the water and the air replace one another, and likewise when other bodies do, so that the place of anything is not a portion or a state of it but is separable from it.

When Aristotle says that places are separable from their occupants, he means that it is possible for something *x* to leave its place and go somewhere else, and for something else *y* to take *x*'s place. However, when *x* moves, it takes its form and its matter with it, because they are logical features of *x*. The argument seems straightforward enough, certainly when applied to Aristotle's candidates for form and matter. It is more difficult to say how it applies to matter conceived as extension. If the extension of something is identified with the particular space it occupies, then presumably the argument fails completely, since the space which something *x* occupies *is* separable from *x* itself. However, if the space which *x* occupies is actually logically constitutive of *x* (as it must be on the extension-as-matter view), then *x* will not be able to give up the space it occupies, on pain of losing its identity, and hence motion will be logically impossible. Moreover, when a certain quantity of water is replaced by a certain quantity of air, we shall be forced to say (on the matter theory of place) that water has *become* air. We shall be unable to distinguish in our philosophical description between the case where this genuinely happens (e.g. when the bath water evaporates) and where water gives way to air (e.g. when the bath is emptied); or at least, we shall not be able to distinguish the two

[67] *Phys.* IV 2, 209^{b}22–8: *τὸ μὲν γὰρ εἶδος καὶ ἡ ὕλη οὐ χωρίζεται τοῦ πράγματος, τὸν δὲ τόπον ἐνδέχεται· ἐν ᾧ γὰρ ἀὴρ ἦν, ἐν τούτῳ πάλιν ὕδωρ, ὥσπερ ἔφαμεν, γίγνεται, ἀντιμεθισταμένων ἀλλήλοις τοῦ τε ὕδατος καὶ τοῦ ἀέρος, καὶ τῶν ἄλλων σωμάτων ὁμοίως, ὥστε οὔτε μόριον οὔθ᾽ ἕξις ἀλλὰ χωριστὸς ὁ τόπος ἑκάστου ἐστι.*

cases by saying that in the one case the water *moves*, and in the other the water *changes into something else*. As Aristotle underlines in his recapitulation of this argument later on, we must be able to say that the water has gone, and the air has come in.[68] 'Water has become air' is not an adequate description of what has happened.

Aristotle's argument against the identification of *x*'s form with *x*'s place, namely that when *x* moves it takes its form with it, is simple and effective, and works if form is conceived of as belonging to the object of which it is the form (as on Aristotle's conception of form). However, just as in the case of matter, the argument may be *too* simple when dealing with the different notion of form implicit in the theory which identifies the matter of something as its extension. The form of the thing, on this theory, is presumably the shape of the object, and is identified with its boundaries. But it is not *wedded* to the particular object of which it is the form, since other properties could come to be in the same space, to yield a different body, and yet one with the same shape. As an account of how reality operates, this is prima facie quite attractive: it is clear, for instance, that when water leaves a jug and air takes its place, the air must take the same form or shape as the water which was in there beforehand. The jug as it were *imposes* a certain shape on its contents. Now, Aristotle is fond of using this particular case as a paradigm case of replacement, and replacement is in turn his model for motion in general. And indeed one might think that when I move through this room the air which takes my place will necessarily have the same shape as I do.

What, then, should Aristotle's more considered objection to this theory be? Presumably it will be along the same lines as his rejection of the matter-as-extension theory. When air replaces water, the same form (in conjunction with the same matter) discards some properties and takes on others; it will therefore be impossible to distinguish between cases where the water *becomes* air and where the water *makes way for* air. What is needed to account for replacement (and therefore, according to Aristotle, motion in general) is a notion of place which allows for bodies to change place while retaining their identity. This is not possible according to both the form-as-place theory and the matter-as-place theory.

It comes as no surprise that Aristotle should refute the identification of place with form and matter by adverting to replacement. It was with that phenomenon that he began his account of place. He

68 IV 4, 211^{b}31–6.

returns to it again when refuting the third and final theory of place which he considers before giving his own account. The reason for this is that Aristotle regards what happens when water is replaced by air in a jug (when I have a jug of water, and I pour the water out) as a model of what happens when any locomotion takes place.[69] Given that a place must be relevantly like a container to allow the analogy to work, Aristotle can rather easily refute the identification of place with matter and form, by concentrating on the case of replacement. Both the identification of place with matter and the identification of place with form, even when these notions are not taken in Aristotle's way, cannot account for the facts of replacement and explain how it is that air takes the place of water as opposed to water becoming air. This is obviously due to the fact that motion is demoted to a *derivative* phenomenon on those views, and no longer is the paradigm change.[70]

Space

Physics IV 2 establishes that the place of x cannot be a logical part of x, that is, that it can be neither x's form nor x's matter. This leaves the third of Aristotle's four candidate definitions left, which states that there is some *interval* which an object occupies which exists over and above that object. Of the four putative definitions of what a place is, this one gets the closest to picking out what we might call 'space'. However, it must not be confused with Plato's conception of *chōra*, even if we translated that word by 'space': Plato's *chōra* is indissociable from a body, for it is its extension, whereas the independently existing interval under discussion in this theory is

[69] See Chapter 1, pp. 23–4. It is ironic that Plato also espouses ἀντιπερίστασις as his account of motion (*Tim.* 79 A 4–81 B 4), despite the fact, as Aristotle shows, that Plato must have a very curious account of what actually happens when something replaces something else.

[70] Aristotle produces other arguments (209^b31–210^a13) to show that the form of x and the matter of x cannot be the place of x. They are as follows: the place of x surrounds x, whereas the matter of x does not (209^b31–2); matter and form do not admit of up and down, so it will be impossible to make sense of the phenomenon of things moving naturally to their own places (210^a2–5); the matter and form of x move with x, and so you will have a moving place, but this means that there must be a place of a place, and this is impossible (210^a5–9); when water comes to be from air, the place of the air should be destroyed since there will no longer be either the form or matter of air, but it makes no sense to say that a place is destroyed (this is an obscure argument, but it must be roughly the sense of 210^a10–13).

no part of the body which occupies it. I shall call this theory of place the *diastēma* theory, because the word Aristotle uses for this separable interval occupied by a body is *diastēma*.

Aristotle introduces and rejects this theory in one of the most difficult passages of *Physics* IV 1–5:[71]

(i) Because it is often the case that the thing which is contained and divided off moves while the container remains still, e.g. [when] water [goes out] of the vessel, what is in between seems to be some sort of extension, as if it were something over and above the body which moves.

(ii) But this is not the case; rather, whichever body it happens to be of those which are moving and which are by nature in contact, falls into place in[72] [the vessel].

(iii) But if there were some extension existing naturally in its own right, and which stays still, there would be an infinity of places in the same spot,

(iv) for as the water and air change position, all the parts will do the same thing in the whole as every bit of water in the vessel:

(v) and at the same time the place will be changing: therefore, there will be some other place than the place, and many places will be coincident.

(vi) But the place of the part, in which it moves, whenever the whole vessel moves, is not some other [place], but the same [as the one of the whole]:

(vii) for air and the water (or the parts of the water) replace each other in the thing in which they are, but not in the place in which they come to be, which is part of the place of the whole universe.

These lines are almost certainly corrupt in several places. The wording of (ii) is slightly awkward;[73] what I have translated as the opening of (iii) does not follow the transmitted text, but rather an emendation based on Themistius' paraphrase[74] of the *Physics*

[71] *Phys.* IV 4, 211^{b}14–29: *διὰ δὲ τὸ μεταβάλλειν πολλάκις μένοντος τοῦ περιέχοντος τὸ περιεχόμενον καὶ διῃρημένον, οἷον ἐξ ἀγγείου ὕδωρ, τὸ μεταξὺ εἶναί τι δοκεῖ, διάστημα ὡς ὄν τι παρὰ τὸ σῶμα τὸ μεθιστάμενον. τὸ δ' οὐκ ἔστιν, ἀλλὰ τὸ τυχὸν ἐμπίπτει σῶμα τῶν μεθισταμένων καὶ ἅπτεσθαι πεφυκότων. εἰ δ' ἦν τι [τὸ] διάστημα ⟨καθ' αὑτὸ⟩ πεφυκὸς ⟨εἶναι⟩ καὶ μένον, ἐν τῷ αὐτῷ ἄπειροι ἂν ἦσαν τόποι (μεθισταμένου γὰρ τοῦ ὕδατος καὶ τοῦ ἀέρος ταὐτὸ ποιήσει τὰ μόρια πάντα ἐν τῷ ὅλῳ ὅπερ ἅπαν τὸ ὕδωρ ἐν τῷ ἀγγείῳ)· ἅμα δὲ καὶ ὁ τόπος ἔσται μεταβάλλων· ὥστ' ἔσται τοῦ τόπου τ' ἄλλος τόπος, καὶ πολλοὶ τόποι ἅμα ἔσονται. οὐκ ἔστι δὲ ἄλλος ὁ τόπος τοῦ μορίου, ἐν ᾧ κινεῖται, ὅταν ὅλον τὸ ἀγγεῖον μεθίστηται, ἀλλ' ὁ αὐτός· ἐν ᾧ γὰρ ἔστιν, ἀντιμεθίσταται ὁ ἀὴρ καὶ τὸ ὕδωρ ἢ τὰ μόρια τοῦ ὕδατος, ἀλλ' οὐκ ἐν ᾧ γίγνονται τόπῳ, ὅς μέρος ἐστὶ τοῦ τόπου ὅς ἐστι τόπος ὅλου τοῦ οὐρανοῦ.*

[72] For this meaning of *ἐμπίπτω* see Hippocrates *De articulis*, § 8, ad init.

[73] 211^{b}18–19.

[74] Dating from the 4th cent. AD.

and first urged by Laas;[75] the repetition of the word *hama* with two entirely different senses in (v) is difficult;[76] it is not clear whether the undesirable conclusion of the view in question is that there is an *infinity* of places in the same spot[77] or that there are *many* places which are coincident.[78] Lastly, and most confusingly, it is not clear whether Aristotle is imagining a case where a vessel is moving—in (i), and implicitly in (iii), it appears to be still, but in (vi) it appears to be on the move.

None the less, I think it is possible to suggest a reasonably plausible interpretation. Let us start by examining what Aristotle says when he sets the problem up. We are to imagine a container which is at rest (*menontos*, 15).[79] Its contents change (*metaballein*, 14–15); Aristotle concentrates on the case of water leaving a vessel (16),[80] but the situation is under-described: *how* exactly is the water leaving the vessel? Perhaps the vessel is being tipped up; perhaps it has had its plug pulled out; or the water is evaporating; or the vessel has been pierced. Or perhaps, quite simply, someone is ladling the water out of the vessel, with a cup or bucket or whatever. Since it is not clear whether the vessel is actually moving or not, it is difficult to choose between all of these, so for the moment let us follow Aristotle's apparent suggestion that the water is leaving the vessel even though the vessel remains still.

In such cases the interval between the walls of the vessel has seemed to some people to be something, namely some sort of interval (*diastēma*) existing over and above[81] the body which actually takes the place of the water. In other words, when a jug is emptied of water, one might reason that there is a gap between the sides of the jug which was *previously* occupied by water, but *now* is not. I take it that the view Aristotle wishes to attack is that this *diastēma* is a *self-subsistent* item; he is *not* denying that a vessel has an interval or gap in which its contents go, but he *is* denying that this interval is an entity in its own right.[82] His view is that the *diastēma* is some *feature* of the contents of a vessel, namely

[75] 211^{b}19–20. See Ross [3], 572–3.

[76] 211^{b}23–5.

[77] 211^{b}20–1.

[78] 211^{b}24–5.

[79] In the following lines numbers such as '15' mean '211^{b}15' etc.

[80] Cf. *Phys.* IV 1, 208^{b}2–3.

[81] Cf. *Phys.* IV 4, 211^{b}9 and 17: *παρά*.

[82] Ross's OCT reads *τὸ μεταξὺ εἶναί τι δοκεῖ διάστημα, ὡς ὄν τι παρὰ τὸ σῶμα τὸ μεθιστάμενον* at 211^{b}17. Hussey [5], 208, suggests *τὸ μεταξὺ εἶναί τι δοκεῖ ὡς ὂν διάστημά τι παρὰ τὸ σῶμα τὸ μεθιστάμενον*, which brings out the intended meaning that according to the *diastēma* theory, what lies between the walls of the vessel 'is

its magnitude,[83] or in other words, that questions about the *diastēma* of a vessel x translate reductively into questions about the dimensions (the volume and presumably shape) of whatever x contains.

Aristotle continues (19) by saying that if there *were* a *diastēma* which existed in its own right[84] and stayed static while the contents of the vessel change over, then there would be an infinity of places in the same spot. The next sentence (bracketed in Ross's edition) is supposed to explain why this conclusion follows (cf. *gar*, 21): it states, apparently, that as the water leaves the jug, it is replaced by air; but the *parts* of the water also leave the jug and are replaced by parts of the final quantity of air just as the *whole* quantity of water leaves the jug and is replaced by the whole of the final quantity of air. Quite how this observation shows that there are infinitely many places in the same spot is not clear, but I suggest the following argument. The *diastēma* theory entails that at every moment t between t_1 and t_2 (where t_1 is the time at which the vessel starts to empty[85] and t_2 is the time at which it has just become empty), there is a different interval—or place, according to this theory—occupied by the water left in the vessel. But the interval occupied by the water in the vessel at any point between t_1 and t_2 is also occupied at t_1 by some water, namely a part of the total water in the vessel at that time, and since the motions of the water as it leaves the vessel and

something'. However, Ross's text is merely an awkward way of expressing something very similar. (Perhaps the comma should be placed after *δοκεῖ* rather than *διάστημα*.)

[83] See e.g. *Phys.* IV 4, 211^{b}8–9, where Aristotle states that the *diastēma* theory is to be rejected if there is no *diastēma* over and above the magnitude of the incoming body (*εἰ μὴ ἔστι μηδὲν διάστημα παρὰ τὸ τοῦ ἐγγιγνομένου σώματος μέγεθος*). Slightly later in *Phys.* IV 4 he explains why some people wrongly think this (212^{a}10–12): *ἐνδέχεσθαι γὰρ φαίνεται εἶναι διάστημα μεταξὺ ἄλλο τι τῶν κινουμένων μεγεθῶν. συμβάλλεται δέ τι καὶ ὁ ἀὴρ δοκῶν ἀσώματος εἶναι* ('It appears to be possible that there is an extension in between, which is something other than the magnitudes which move. Air, too, helps somewhat to give this impression by appearing to be incorporeal').

[84] Note that Ross's text at 19–20 (*εἰ δ' ἦν τι [τὸ] διάστημα ⟨καθ' αὑ⟩τὸ πεφυκὸς ⟨εἶναι⟩ καὶ μένον*), following a conjecture by Laas, merely underlines the supposition which is in any case implicit in Aristotle's attack, that the *diastēma* theory of place fails to acknowledge that, properly understood, the *diastēma* of a vessel is not an entity in its own right, but reduces to features of bodies. Even if one does not adopt that text, this is how Aristotle must be understood. However, there is good support for the addition of *καθ' αὑτό* from *Phys.* IV 8, 216^{a}24, where Aristotle attacks the similar notion of an independently existing void which is *ἀποκεκριμένον καθ' αὑτό*.

[85] I ignore the fact that according to Aristotle, there is no such time—or at least, no such moment (cf. *Phys.* VI 5). The argument will still go through if t_1 is taken to be some moment after the start of the emptying of the vessel.

of the air as it replaces the water are infinitely divisible, there will be an infinity of *overlapping* or *coincident* places occupied by water at t_1. The same argument can be constructed for air: at t_2 there are an infinite number of coincident (overlapping) intervals occupied by air.

In effect, this argument suggests that the water in a vessel occupies a particular interval, but that there are infinitely many other intervals *within* this interval,[86] each of which will be occupied at some point during the water's departure from the vessel by some incoming air. At any time during the evacuation of the water, the question 'where was the water which has now left the vessel?' has a simple answer—it occupied the interval which has now been taken up by the air in the vessel. These intervals or places (of which there are infinitely many) are all occupied at the beginning of the process by water, and all occupied at the end of the process by air. But they are all coincident or overlapping.

That this is roughly what Aristotle intends to say is confirmed by what he says in order to *block* the argument: 'But the place of the part, in which it moves, whenever the whole vessel moves, is not some other [place], but the same [as the one of the whole]' (25–6).[87] The point of this objection is that, for instance, at t_1 it would not be right to say that a *part* of the total water occupies a place different from that of the total water (nor at t_2 would it be right to say that a part of the air in the vessel occupies some other place than the place of the total air). In other words, the principle to which the *diastēma* theory commits you (namely, that some or all of the *parts* of the water in a vessel occupy places), and which is responsible for generating the unacceptable argument that there are infinitely many places overlapping, is false. Aristotle had in fact already observed in 211^{a}29–^{b}1 that the parts of a continuous whole do not relate to their immediate surroundings in the same way as something separate and contained does; he said that when something is continuous with its surroundings, it is not in them *as in a place* but rather *as a part is in the whole*, so that it does not move *in* its surroundings but rather *with* them. We are reminded that a part of a continuous whole does

[86] Cf. Ross [3], 573: 'an infinite number of self-subsistent intervals, one inside another like the boxes in a nest of boxes'.

[87] 211^{b}25–6: *οὐκ ἔστι δὲ ἄλλος ὁ τόπος τοῦ μορίου, ἐν ᾧ κινεῖται, ὅταν ὅλον τὸ ἀγγεῖον μεθίστηται, ἀλλ' ὁ αὐτός*. I shall come back to the troublesome reference to the vessel moving.

not move *in* something in 25–6,[88] so in the case of the vessel being emptied of water, the parts of water do not actually *have* a place at t_1, for they were continuous with the water and hence did not have a place different from the place of the whole water.

On the correct view, there are just bodies of water and air in the vessel, which occupy different places as they change in size. This is presumably what Aristotle rather awkwardly tries to express in the next sentence: 'for air and the water (or the parts of the water) replace each other in the thing in which they are, but not in the place in which they come to be, which is part of the place of the whole universe'.[89] In other words, there is no need to posit an *extra* place for the parts of water and air, e.g. a place which would be some part of the *diastēma* of the whole universe. (I shall be coming back to this last point.)

As I have interpreted it, the puzzle bears some similarity to the sixth *aporia* of *Physics* IV 1, where Aristotle considers the problem of a body which is growing and wonders whether its place also grows with it: 'Again, just as every body is in a place, so is there a body in every place. What, then, are we to say about things which increase in size? For it is necessary from these [premisses] that their place increase with them, if the place of each thing is not smaller or larger than it.'[90] The puzzle is a good one, and stems from the fact that one can say in Greek as in English 'my weight increased', 'the number of people in the room changed constantly' etc. This way of speaking should not, however, be taken to imply that weights and numbers themselves are subject to change, any more than a locution of the sort '*x*'s place is changing' implies that places are subject to change.[91] Aristotle's answer to the *aporia* makes this clear: 'For in increase and decrease the thing changes, and what was previously

[88] Cf. 211ᵃ35: a part continuous with its whole *οὐκ ἐν ἐκείνῳ κινεῖται*; 211ᵇ25–6: *οὐκ ἔστι δέ ἄλλος ὁ τόπος τοῦ μορίου, ἐν ᾧ κινεῖται.*

[89] 211ᵇ27–9: *ἐν ᾧ γὰρ ἔστιν, ἀντιμεθίσταται ὁ ἀὴρ καὶ τὸ ὕδωρ ἢ τὰ μόρια τοῦ ὕδατος, ἀλλ' οὐκ ἐν ᾧ γίγνονται τόπῳ, ὃς μέρος ἐστὶ τοῦ τόπου ὅς ἐστι τόπος ὅλου τοῦ οὐρανοῦ.*

[90] *Phys.* IV 1, 209ᵃ26–9: *ἔτι ὥσπερ ἅπαν σῶμα ἐν τόπῳ, οὕτω καὶ ἐν τόπῳ ἅπαντι σῶμα· πῶς οὖν ἐροῦμεν περὶ τῶν αὐξανομένων; ἀνάγκη γὰρ ἐκ τούτων συναύξεσθαι αὐτοῖς τὸν τόπον, εἰ μήτ' ἐλάττων μήτε μείζων ὁ τόπος ἑκάστου.*

[91] Frege's puzzle of the increasing rod is very similar: the number which gives the length of the rod in millimetres is 1000 at one moment, and 1001 at another, but 1000 has not *become* 1001. Cf. Frege [104], 285–6 (original 657–8). Frege insists that we should say 'the number which gives the length of the rod at time t_1 is 1000' and 'the number which gives the length of the rod at time t_2 is 1001'; then we have two different subjects of predication, and cannot draw the inference that something which was 1000 is now 1001.

there has now changed into a smaller or larger [place].'[92] As the object grows, it occupies ever larger places; there is not *one* place which grows with it.

The twist to the puzzle as it appears in the attack on the *diastēma* theory of place is that according to that theory, an increasing body occupies a succession of places each of which *remains* occupied (by a part of the body) while that body increases, thus giving rise at the end of the increase to an infinity of coincident or overlapping places all piled up on top of each other.

So far, I have not discussed one rather puzzling sentence (23–5) in Aristotle's argument, which I translated above as follows: 'and at the same time the place will be changing: therefore, there will be some other place than the place, and many places will be coincident'.[93] My interpretation of these lines is that they say that the place of the ever decreasing body of water in the vessel is changing because the water occupies a succession of different places while it decreases in volume, and that according to the *diastēma* theory there will not be just the place of the water (and only that place) at any time during the evacuation of the vessel, but that there will be other places, namely those corresponding to the parts of the water.

Before discussing one of the other possible interpretations of these lines, I want to point out that it is clear that the next sentence, 'But the place of the part, in which it moves, whenever the whole vessel moves, is not some other [place], but the same [as the one of the whole]',[94] is supposed to reply to the faulty consequence that the *diastēma* theory drew in 23–5, that there is an *extra* or *different* place at each instant of the water's replacement of air. I think, therefore, that lines 23–5 should be taken as continuing the explanation which started in lines 21–3 (which are bracketed by Ross) of why there will be on the *diastēma* theory many coincident places.[95]

Of course, lines 23–5 could also be translated: 'but the place will

[92] *Phys.* IV 4, 211^{a}15–17: *καὶ γὰρ ἐν τῇ αὐξήσει καὶ φθίσει μεταβάλλει, καὶ ὃ πρότερον ἦν ἐνταῦθα, πάλιν μεθέστηκεν εἰς ἔλαττον ἢ μεῖζον*. As Hussey [5], 112, points out, we must supply a word meaning something like 'region' with the neuter adjectives in order to obtain something which is appropriate to the context.

[93] *Phys.* IV 4, 211^{b}23–4: *ἅμα δὲ καὶ ὁ τόπος ἔσται μεταβάλλων· ὥστ' ἔσται τοῦ τόπου τ' ἄλλος τόπος, καὶ πολλοὶ τόποι ἅμα ἔσονται.*

[94] *Phys.* IV 4, 211^{b}25–6: *οὐκ ἔστι δὲ ἄλλος τόπος τοῦ μορίου, ἐν ᾧ κινεῖται, ὅταν ὅλον τὸ ἀγγεῖον μεθίστηται, ἀλλ' ὁ αὐτός.*

[95] There is some linguistic evidence for taking them together: the verbs in 23–5 are all in the future (this is the normal future implicative: 'if such and such is the case, then this will be the case'), co-ordinate with *ποιήσει* in line 21. Aristotle switches to

be moving at the same time and so there will be another place of the place, and there will be many places together'. In this case, these lines constitute a subargument against the identification of the interval between the walls of the jug as a place, since Aristotle will then be imagining that the vessel is tipped over and that in consequence the interval between its walls moves; hence there is a place of the interval, but places do not have places.[96]

This interpretation (which I shall call the 'subargument' interpretation) has one major advantage, and several disadvantages. The advantage is that Aristotle refers in the very next sentence to what happens 'when the whole vessel moves', and this suggests that Aristotle had something like the subargument in mind. However, there is a difficulty here. As I have already said, Aristotle's argument in lines 23–5 that there is not 'another place' of any part of the water or air surely takes up his reference in line 24 to there not being 'another place'. But in that case there should be continuity between lines 19–23 and 23–5; they contain a continuous argument treating the supposed consequence of the *diastēma* theory that there will be infinitely many places piled up together, and should not be interrupted by another argument.

This is not the only problem for the subargument interpretation. The *diastēma* theory says that there is such a thing as a self-subsistent interval which bodies occupy. It does *not* follow from this that this interval must be tied to a particular vessel. In fact, in lines 28–9 Aristotle seems to conceive of the *diastēma* which something occupies as being part of the *diastēma* of the whole universe, and this would be inconsistent with the view that a vessel has an independent mobile *diastēma* (which retains its identity as it moves) that the vessel's contents occupy. Furthermore, in *Physics* IV 8, where Aristotle discusses a theory of self-subsistent void very similar to the *diastēma* theory here discussed,[97] he takes that view as imply-

the present in line 25 (*ἔστι*) because he is then telling us what he thinks to be the case. According to my interpretation, Ross's brackets close too early.

[96] For this interpretation see e.g. Ross [3], 573: 'This is the second paradoxical consequence, that when a container moves, the self-subsistent interval which, on the view in question, it has inside it, will be undergoing a change, i.e. will be moving to occupy another interval.'

[97] *Phys.* IV 8, 216^{a}23–^{b}21. Cf. 216^{a}24–6: *τοῦτο δὲ ταὐτόν ἐστι τῷ τὸν τόπον φάναι εἶναί τι κεχωρισμένον· τοῦτο δ' ὅτι ἀδύνατον, εἴρηται πρότερον* ('This is the same as saying that something's place is separated from it: but it has already been said that this is impossible').

ing that bodies occupy parts of void[98]—presumably, the entire void occupied by the universe.

Again, Aristotle's argument against the *diastēma* theory *should* work for an immobile vessel, for he takes replacement to be the model for motion in general (i.e. for displacement), so we should expect whatever arguments he gives to be easily adaptable to the general case where we abstract from the particularities of the situation where there is a vessel in the more literal sense. This cannot be done with the supposed subargument, which relies on the *diastēma* in question being tied to a particular vessel and being mobile.

On my interpretation, therefore, the reference to the moving vessel in line 26 is inappropriate, and I suggest that the phrase was an interpolation, probably by someone who did not understand the meaning of 'at the same time the place will be changing' in 23, and assumed that Aristotle must have meant that the place of the water was *moving*, not *changing*. However, I do not wish to dwell or insist upon my interpretation of these few lines. It is important not to get lost in the details of justifying this or that interpretation, and I shall take it that I have established at any rate that Aristotle's *principal* argument against the *diastēma* theory is that an increasing body will cause places to be 'stacked up' because its parts can be seen to occupy overlapping places.

The *diastēma* theory has an account to give of what happens not only when something replaces something else in a vessel, but also of what happens when something moves quite generally. In other words, it is possible to extend the *diastēma* theory to deal with not only *replacement*, but also any case of locomotion, i.e. *displacement*. Moreover, Aristotle's objection to the *diastēma* theory can also be generalized as an objection to this more general account. The following is one way in which it might be extended.

Suppose something, x, moves from here to the other side of the room. Then take the beginning of this move. x occupies interval i_1 and then slightly later interval i_2. Suppose that in fact 'slightly later' is taken in such a way that i_1 and i_2 overlap; but in that case their common part (i.e. some interval or other) was occupied by *part* of x at t_1. Hence, by taking sufficiently small intervals of time, you can construct arguments to show that at t_1 an infinity of different intervals were occupied by parts of x. Hence the place x occupies

[98] *Phys.* IV 8, 216^{b}8–9: ἐν τῷ αὐτῷ ἔσται τῷ τοῦ τόπου καὶ τῷ τοῦ κενοῦ μέρει ἴσῳ ἑαυτῷ ('It will be in the same part of place, i.e. of void, equal in magnitude to itself').

at t_1 is in fact host to an infinity of other places within places etc. The proof is simple and neat.

In summary, the argument of 211ᵇ14–29, as I see it, runs more or less as follows. When air replaces water in the vessel, then at each moment during the replacement the water in the vessel occupies a different, progressively smaller, interval, and the air occupies a different, progressively bigger, interval. These progressively bigger intervals occupied by the air are in fact all occupied by parts of the air at the end of the emptying of the vessel; similarly, the progressively smaller intervals occupied by the water were in fact all occupied by parts of the water at the beginning of the process. Hence, there are infinitely many intervals all within the interval occupied by the total air which has replaced the water (or, equivalently, an infinity of intervals within the interval occupied by the total water). Aristotle describes this as a situation where there are an infinity of places in the same spot (211ᵇ20–1), or (ironically, I suppose) *many* places coinciding (211ᵇ24–5).

Aristotle's argument makes use of the notion of being *coincident* (*hama*), whereas in my interpretation of his argument I have used the notion of *overlapping*. In fact, it is quite natural to describe two things as 'in the same spot' even if they are only *partially* in the same spot. For instance, at *Physics* IV 4, 211ᵃ33–4, Aristotle points out that the limits of two things which touch are 'in the same spot'. But clearly, two things side by side do not have *completely* coincident boundaries; even when one thing is nestled inside another, their boundaries are not completely coincident (think of the external boundary of the outer object). So things can be 'in the same spot' even if they are not completely coincident; in other words, even if they overlap. (This is also the case when we talk of temporal coincidence: I arrive at the supermarket; you have already done half your shopping—you finish your shopping and leave before me. Our visits to the supermarket *coincided*.)

There is reason none the less to beware of the term 'overlapping', for there are two types of overlapping. In the first sense, two things overlap if they share a (not necessarily proper) part. For instance, the top two thirds of my body overlap the bottom two thirds of my body, because they share a part (the middle third). In the second sense, two things overlap not in so far as they *share* a part, but in so far as they have *coincident* parts. So the boundaries of two bodies which are in contact overlap in the second way of overlapping: a part of one is coincident with part of the other.

It is the first way of overlapping which is in question in *Physics* IV 4.[99] The infinitely many overlapping intervals are each parts of the whole interval between the walls of the vessel; the air which comes into the vessel will therefore occupy ever larger parts of that interval. But what *exactly* is the unacceptable consequence of this view? At any moment during the emptying of the vessel, the interval occupied by the incoming air is ever greater than before, but we need not view this as *different* intervals all stacking up on top of each other. Suppose that at a given moment the air occupies interval i_1, and then slightly later it occupies interval i_2. The second interval is larger than the first and overlaps it; but why think that these are two *different* intervals one laid over the other? Why not think instead that i_1 is *part* of i_2? What is the metaphysical difficulty with that?

Aristotle does not help us in answering this question, but there is a possible answer. On the *diastēma* theory the *parts* of the water (and the parts of the incoming air) occupy *actually existing* intervals, namely parts of the whole *diastēma* extended between the vessel walls. However, on Aristotle's view, only if a part of something is somehow divided off from its whole does it deserve a *place* to itself, and the *parts* of water or air are therefore disqualified. We end up with an embarrassment of places for items which, so to speak, do not deserve them, being continuous parts of a whole which occupies *one* place. Aristotle thinks that the only places needed to explain or describe adequately what is going on in the case of the replacement of water by air are the place which the water remaining in the vessel holds and that which the incoming air holds. These places change during the emptying, but they do not pile up or hang around, doing no explanatory work. Yet according to the *diastēma* theory, these places are constantly present, occupied by *parts* of the water and air. These places are constantly present precisely because they are

[99] If the intervals in the example of the water and the jug are taken to overlap in the second way, places will be taken to be the sort of thing which can be in the same spot, like boundaries. However, if two things overlap in this second way, then it is because part of one is in the same place as part of the other. But obviously we cannot say this about two intervals which overlap in this way, since their places (if it is even possible to talk of their places) will be nothing other than the intervals *they* occupy, which in turn will overlap in this second way etc. etc. Intervals are, therefore, the sort of thing which, if they overlap, must overlap in the first way.

parts of the whole *diastēma*, and are occupied by parts of the water or air. Yet they are not needed.

Aristotle's argument against the *diastēma* theory of place emerges as a rather laborious but ingenious one. However, there is a rather simpler objection which one might make to that theory, which is itself reminiscent of an objection Aristotle makes in *Physics* IV 8 to the view that there is a self-subsistent *void*.[100] The *diastēma* which is supposed to be the interval occupied by a body is itself something extended. So it must itself occupy an interval. (What else is there to being extended other than occupying a certain interval?) But then that interval which it occupies is itself extended—and so on. This argument establishes a *reductio ad infinitum* which is damaging for the *diastēma* theory in this way: it will be impossible to say where the interval occupied by something is (for similar reasons to those advanced in the discussion of Zeno's paradox of place). But that is absurd: as Aristotle suggested, it *is* possible to specify where the place of something is, as long as this 'where' is understood in the right way. This argument is similar to certain aspects of Aristotle's argument against the *diastēma* theory of place in *Physics* IV 4; for instance, the *conclusion* of that argument, that there are an infinity of places on top of each other, is also one way of stating the conclusion of the argument based on *Physics* IV 8. Perhaps it is worth speculating as to whether *Physics* IV 4, 211^b14–29, did contain such an argument but there has been extensive corruption.

I have argued that the objection in *Physics* IV 4 stems from economy of explanation and metaphysical propriety, for on the *diastēma* theory of place, there are an infinity of places 'hanging around' when water is replaced by air in a vessel, because all the parts of the outgoing water and the incoming air are accorded a place, whereas they are not the right sort of item to have a place. There is another objection implicit in *Physics* IV 8 according to which intervals would have themselves to have places, and so on *ad infinitum*. Either way, Aristotle has some interesting and forceful arguments against the notion of a self-subsistent space.

[100] *Phys.* IV 8, 216^b10–11: *καὶ εἰ δύο τοιαῦτα, διὰ τί οὐ καὶ ὁποσαοῦν ἐν τῷ αὐτῷ ἔσται;*

5

Aristotle's Concept of Place

Introduction

As seen in the previous chapter, the place of something, *x*, is not *x*'s form or *x*'s matter, nor is it the interval between the limits of what contains *x*. But this leaves the fourth of those possibilities for what a place is which Aristotle distinguished in *Physics* IV 4, 211^{b}6–9,[1] namely 'the limit of the surrounding body ⟨at which it is in contact with that which is surrounded⟩'.[2] This is a definition of *x*'s *proper* place, i.e. that in which *x* is *non-derivatively* (so much is clear from the fact that the definition clearly means to pick out that place of something which is the same size as it). 'In' here has the same sense as it does when we say that something is 'in a place' (*en topōi*), that is, the 'locative' sense of 'in'—the eighth way in which something is said to be in something else.[3] The compound of *echein* ('to have') which corresponds to this sense of 'in' is *periechein*, which means 'to surround';[4] hence Aristotle said that your proper place 'surrounds no more than you'.[5]

It is not surprising that Aristotle should attempt a definition of *x*'s proper place, since proper places are the key to his more general

[1] See Chapter 4, pp. 103–5, for this division.

[2] IV 4, 212^{a}6–6*a*: τὸ πέρας τοῦ περιέχοντος σώματος ⟨καθ' ὃ συνάπτει τῷ περιεχομένῳ⟩. See Ross's note ad loc. for the addition of καθ' ὃ συνάπτει τῷ περιεχομένῳ, following Simplicius, Philoponus, Themistius, and the Arabo-Latin translation. In his final refined definition of what a place is Aristotle will say that it is the *first* (πρῶτον) immobile limit of what surrounds something, and Simplicius actually *says* that Aristotle uses the word πρῶτον in the second definition as shorthand for the phrase καθ' ὃ συνάπτει τῷ περιεχομένῳ in the first (*In Phys.* 584. 19–20). Something which surrounds me standardly has at least two limits; one where it is in contact with me (the 'first' limit of what contains me), and one where it is in contact with whatever is outside it (which could presumably be called the 'last' or 'outer' limit). Note that the subject of συνάπτει is 'the surrounding body' (and not 'the place', *pace* Lang [48], 92–3).

[3] For this see Chapter 2, pp. 71–6.

[4] Cf. *Meta.* Δ 23, 1023^{a}13–14: τὸ περιέχον, τὰ περιεχόμενα, περιέχοντι. See Chapter 2, pp. 76–8.

[5] *Phys.* IV 2, 209^{b}1: ὃς περιέχει οὐδὲν πλέον ἢ σε.

account of the notion of being somewhere. They are the key in this sense: everything is somewhere, but some things are somewhere in virtue of being in a place (they are P-items, i.e. items with a place); others are somewhere in virtue of being in P-items (in some other sense of 'in');[6] therefore P-items provide the home for the other things which are somewhere; but it is in virtue of being in their *proper* places that these items are in *any* of their places.[7] Or, if you prefer: P-items have a unique place (called its proper place on the other conception), which can be specified more or less precisely by adverting to different items (broad places).

In *Physics* IV 2 the proper place of something appeared to be the most precise place in which it is. Aristotle claimed that this place would inevitably be some sort of limit (*peras*). This appeared plausible in the case of something entirely surrounded by (for instance) air, but it was not clear what to say in more complex cases—for instance, where something is on the earth, in contact with the earth and the air, and perhaps other things too. I hope to show that even in such complex cases it is possible to make sense of what the limit of something's surroundings is.

Even then, there is an important final step in Aristotle's account. After arguing that the place of something is the limit of the surrounding body at which it is in contact with that which is surrounded, Aristotle reformulates his definition a few lines later as 'the first immobile limit of that which surrounds, that is the place of something'.[8] Aristotle introduces the qualification 'immobile' (*akinēton*) as a refinement to his definition. In this chapter I shall tie up the loose ends of my presentation so far of Aristotle's concept of place, and then tackle the considerations which lead to the introduction of the qualification 'immobile'.

The body which surrounds

Throughout *Physics* IV 1–5, Aristotle often refers to *to periechon*, 'that which surrounds', with or without the substantive 'body'

[6] See Chapter 3, pp. 97–8.

[7] Predicates of the form '. . . is in such-and-such' (where the 'in' is taken in the locative sense) hold of *x* in virtue of the fact that *x* is in its proper place. See Chapter 2, pp. 78–80.

[8] *Phys.* IV 4, 212ᵃ20–1: *τὸ τοῦ περιέχοντος πέρας ἀκίνητον πρῶτον, τοῦτ᾽ ἔστιν ὁ τόπος.*

(*sōma*). One of the reasons why Aristotle never defines this notion is because *to periechon* is a familiar term in Presocratic cosmology.[9] For instance, Anaximander is reported as thinking that the infinite surrounds the heavens,[10] and Anaxagoras appears to have thought the same: 'For air and ether are separating off from the surrounding mass. And what surrounds is infinite in quantity.'[11] Another important passage is from Aristotle's *Physics*:[12]

> And so, as I say, the infinite is thought not to have a principle [to which it is subject] but itself to be a principle of other things, and to 'surround everything' and 'steer everything' as is said by all those [natural philosophers] who do not provide other explanatory factors besides the infinite.

Something's surroundings envelop it, and are in contact with it; they could even be considered to govern the thing, or hold it in:[13] Empedocles thought that the sun was stopped from going in a straight line by the sphere surrounding it,[14] and Aristotle reports in the *De anima* how earlier thinkers thought that something's surroundings (*to periechon* again) pressed in on it, causing respiration.[15] Even Aristotle himself argues that some of the motions produced in an organism are perhaps due to its surroundings (*to periechon* once more).[16] Since Aristotle believes that something can only have an effect on something else if it is in contact with it, this again indicates that *to periechon* is in contact with what it surrounds.

The verb *periechein* was no stranger to ancient discussions of what places are. In the *Parmenides* the question arises: is the One somewhere? Being somewhere is identified with being in something, and Plato has Parmenides show first that the One cannot be in some-

[9] See DK 'Wortindex', s.v. περιέχειν.

[10] 12 A 11 DK: πάντας περιέχειν τοὺς κόσμους. See Kahn [29].

[11] 59 B 2 DK: καὶ γὰρ ἀήρ τε καὶ αἰθὴρ ἀποκρίνονται ἀπὸ τοῦ πολλοῦ τοῦ περιέχοντος, καὶ τό γε περιέχον ἄπειρόν ἐστι τὸ πλῆθος. Translation from Barnes [9], 227.

[12] *Phys.* III 4, 203^b10–13 = 12 A 15 DK: διό, καθάπερ λέγομεν, οὐ ταύτης ἀρχή, ἀλλ' αὕτη τῶν ἄλλων εἶναι δοκεῖ καὶ περιέχειν ἅπαντα καὶ πάντα κυβερνᾶν, ὥς φασιν ὅσοι μὴ ποιοῦσι παρὰ τὸ ἄπειρον ἄλλας αἰτίας. Translation from Hussey [5].

[13] It is interesting to recall that in one of the other senses of ἐν (the sixth) '*x* is ἐν *y*' means that *y* governs *x*; recall also that ἔχειν can mean 'have a hold over'. (See Chapter 2, p. 76.) In a certain sense, a container does 'master' its contents, because it holds them in and stops them from spilling out.

[14] 31 A 58 DK: τῆς περιεχούσης αὐτὸν σφαίρας.

[15] *DA* I 2, 404^a10 = 67 A 28 DK: συνάγοντος . . . τοῦ περιέχοντος τὰ σώματα.

[16] *Phys.* VIII 2, 253^a12–13: τούτου δὲ τῆς κινήσεως οὐκ αὐτὸ τὸ ζῷον αἴτιον, ἀλλὰ τὸ περιέχον ἴσως.

thing else, and then that it cannot be in itself. He says:[17] 'If it were in another, it would be encompassed all round by that in which it was, and would have many with it at many points; but there cannot be contact at many points all round with a thing which is one and has no parts and is not round.' In the *Parmenides*, therefore, being surrounded by something else, in such a way that being in that thing counts as being somewhere, entails being in contact with it on all sides.

Aristotle has in mind this sort of surrounding—what I termed 'proximate surrounding'—when he offers his definition of place. This is particularly visible when he says that the place of something is 'the limit of the surrounding body ⟨at which it is in contact with that which is surrounded⟩'.[18] So 'surrounding' should not be taken in the sense of 'surrounding' in which I am surrounded by the celestial sphere—remote containing—but rather that in which a fish is surrounded by water. Perhaps the English 'surroundings'[19] captures this idea; the Oxford translator of the *De anima* has 'the environment'.[20]

So much for 'surrounding'. But when Aristotle defines a place as 'the limit of the surrounding body ⟨at which it is in contact with that which is surrounded⟩', what does 'body' mean? A body is anything three-dimensionally extended; that is, with length, breadth, and depth.[21] Elsewhere, Aristotle says that a body is what is bounded by a surface.[22] There is no reason to confine ourselves to substances: Aristotle refers to the four elements as 'the natural simple bodies' and he does not think that they are substances. Indeed, they are just heaps: in *Metaphysics Z* 16 and 17 he distinguishes between assemblages of items which form a unity (and hence are substances) and those which do not (which are just heaps, like the elements).[23]

[17] *Parm.* 138 A 3–7: *ἐν ἄλλῳ μὲν ὂν κύκλῳ που ἂν περιέχοιτο ὑπ' ἐκείνου ἐν ᾧ ἐνείη, καὶ πολλαχοῦ ἂν αὐτοῦ ἅπτοιτο πολλοῖς· τοῦ δὲ ἑνός τε καὶ ἀμεροῦς καὶ κύκλου μὴ μετέχοντος ἀδύνατον πολλαχῇ κύκλῳ ἅπτεσθαι.* Translation from Cornford [10], 119.

[18] *Phys.* IV 4, 212ᵃ6–6ᵃ. See n. 2 above for the bracketed words.

[19] *OED*: 'sum total of things or conditions in the neighbourhood of a person or thing'.

[20] See [1], translating *DA* I 2, 404ᵃ10.

[21] Cf. *Top.* VI 5, 142ᵇ24–5: *ὁ τοῦ σώματος ὁρισμὸς τὸ ἔχον τρεῖς διαστάσεις* with *Phys.* IV 1, 209ᵃ4–6: *μῆκος καὶ πλάτος καὶ βάθος, οἷς ὁρίζεται σῶμα πᾶν.* Cf. Euclid *Elements* XI, def. 1: *στερεόν ἐστι τὸ μῆκος καὶ πλάτος καὶ βάθος ἔχον.* See Heath [14], iii. 262–3.

[22] *Phys.* III 5, 204ᵇ5–6: *τὸ ἐπιπέδῳ ὡρισμένον.* I discuss the meaning of *ἐπίπεδον* below, pp. 140–1.

[23] *Meta.* *Z* 16, 1040ᵇ8–10: *οὐδὲν γὰρ αὐτῶν* [parts of animals, elements] *ἕν ἐστιν, ἀλλ' οἷον σωρός, πρὶν ἢ πεφθῇ καὶ γένηταί τι ἐξ αὐτῶν ἕν.*

Heaps count as bodies, as do unified heaps, as it were, which are substances.

Now, when Aristotle talks of 'the surrounding body', he uses the definite article—he appears to have just *one* such body in mind. But often there is no single simple thing which surrounds something. For instance, if I am standing up in the shallow end of a swimming pool, I am (if all is well) surrounded by no single thing; rather, I am partly surrounded by air, partly by water—or again, most of the time in everyday life I am surrounded by (and in contact with) air and clothing. Aristotle has insisted throughout on the analogy between motion and the replacement of water by air in a vessel, but even the water in a jug is surrounded by a combination of the vessel and the air with which the top of the water is in contact. Neither the vessel nor the air, taken singly, surrounds the water completely, but the water is completely surrounded by the vessel and the air taken together. In other words, the water's immediate surroundings consist of the vessel plus some air. The proximate container of *x* can thus be a *collection* or *assemblage* of items.[24] (So the English expressions '*x*'s surroundings', or 'what surrounds *x*', convey the idea better than 'the body which surrounds *x*'.)

There are different ways of referring to something's surroundings, depending on quite how much we take in. For instance, a fish in the sea off the west coast of Greece is surrounded by (and in contact with) water, the sea, the Mediterranean Sea, the Ionian Sea, etc. Although these different specifications of the fish's surroundings do not pick out the same body, it does not seem right to say that they pick out completely different bodies either. This is, of course, extremely puzzling—even paradoxical. Yet it was a familiar idea in ancient philosophy, starting perhaps with Plato, and finding its best friends among the Stoics.[25] 'In general, if the Stoics are right, it is possible for there to be two different things, X and Y,

[24] Barnes points out that ancient philosophers are reluctant to talk of collections of things without 'a structure and a character of their own' ([88], 249) as *wholes*. Doubtless, this is why Aristotle says that a *vessel* surrounds its contents, the *air* surrounds me, and the *river* surrounds the boat, despite the fact that, strictly speaking, they do not surround *completely*. He goes for the wholes. Aristotle's concept of place, I think, needs first to be elucidated strictly (in a way which would be appropriate to underpin a scientific account of place); then it can be relaxed to include 'non-strict'—but whole—surrounders, and so account for our more loose way of talking in everyday life. See below, pp. 161–4, for the relaxation.

[25] See Barnes, [88], 262–8.

both of which are Fs—and yet for there to be only *one* F. Perhaps the view *is* paradoxical. But it is also true.'[26] Aristotle appears to go along with the view too, for there are a range of non-identical things, all of which are bodies surrounding *x*, and yet Aristotle constantly talks of *the* body surrounding *x*. The explanation of this apparent paradox is clearly enunciated by Seneca and attributed to the Stoics: 'One thing must be separate from another if they are to be two.'[27] In the example of the fish in water, the different bodies surrounding the fish are not in fact separate from one another. The fish is surrounded by the sea, *and* the Mediterranean: but the Mediterranean is *part* of the sea (considered as the whole of the sea which surrounds the globe); similarly, the Ionian Sea is *part* of the Mediterranean.

But why stop there? Why not simply say that the fish is surrounded by water? Not just sea water (for our purposes, a species of water) but water, the element (the sum of all the water that there is, what Quine calls a 'scattered object').[28] Clearly, all the other bodies mentioned are just parts of water, in this wide sense. But why not continue, and take the sum of water and air? (After all, assemblages, or fusions, have been admitted into the scheme already.) Anything surrounded by water is surrounded by the sum of water and air—trivially. This summing has an upper limit, for according to Aristotle the universe is finite. So *x* is surrounded by the universe—as is everything except the universe itself, which is why Aristotle calls it the 'common place of all things'.[29] Aristotle says in *De caelo* I 9 that the universe is the body (*sōma*) which is the sum of all things.[30] The universe is precisely fitted to *x*, since there is no void, and is '*x*'s maximal surrounder'.[31]

So it is not the case that *x* is surrounded by many *separate* bodies. The reason why they are not separate is because there is a body which surrounds *x* such that all the other bodies which surround *x* are *parts* of it: this body is *x*'s maximal surrounder, and holds title

[26] Ibid. 265 (his italics). Note that we cannot say that X and Y are the same F—even though there is only one F.

[27] *Ep.* CXIII 4 (cited by Barnes [88], 264). The word for 'separate' is *diductus*.

[28] See Quine [126], 98–9. If the element water is a scattered entity—and a body—then it has length, breadth, and depth; it is not easy to say *what* length, breadth, and depth.

[29] *Phys.* IV 2, 209^{a}32. See Chapter 2, pp. 55–66.

[30] *De caelo* I 9, 278^{b}20.

[31] The maximal surrounder of something, *x*, should not be defined as the sum of everything except *x*. I explain why below, p. 147.

to being *the* body which surrounds *x*. When Aristotle talks of *the* body which surrounds *x*, he does not mean that we must pick some simple body in the immediate surroundings of *x*; rather, I suggest, he refers to the maximal surrounder of *x*.

The limit at which it is in contact with what it surrounds

Something, *x*, may have many surrounding bodies, but this does not mean that it has many *different* surrounding bodies. They are not distinct; they are all parts of the maximal proximate surrounder of *x*. However, it is not the case that every part of *x*'s maximal surrounder proximately surrounds *x*. For instance, one part of my maximal surrounder is the sphere of fire, but (luckily) the sphere of fire does not proximately surround me. What condition does a part of the maximal surrounder of *x* have to fulfil in order proximately to surround *x*? The answer is: it must include the limit of the maximal surrounder at which it is in contact with *x*. Any part which includes this limit proximately surrounds *x*. (So take the fish submerged in the sea: any body—including assemblages—which includes the surface of water surrounding the fish proximately surrounds the fish.) It is in virtue of possessing this limit that anything counts as surrounding *x*. So it is this surface which is the thing which surrounds *x* ***kath' hauto*** or non-derivatively—it is this surface which is the item in which *x* is ***kath' hauto***, in the locative sense of 'in'.[32] Thus, the limit at which *x* is in contact with its surroundings is of some importance, for it determines what things will count as surrounding *x*.

The central question here is to justify talking of the limit (*peras*) of the maximal surrounder. Many different things have limits. In geometry lines have limits, planes have limits, and solids have limits. (In fact, points are the limits of lines; lines are the limits of planes; planes or surfaces are the limits of solids.) But the maximal surrounder of something is a body. What is the limit of a body? A limit is what a body has if it is not infinite; the Greek word for 'infinite' (*apeiron*) means literally 'unlimited', without a *peras*.[33] Aristotle remarks on this in *Metaphysics Δ*:[34] 'We call a limit the

[32] See Chapter 2, pp. 62–3.

[33] Cf. *Phys.* III 5, 204^{b}5–6.

[34] *Meta.* *Δ* 17, 1022^{a}4–5: *πέρας λέγεται τό τε ἔσχατον ἑκάστου καὶ οὗ ἔξω μηδὲν ἔστι λαβεῖν πρώτου καὶ οὗ ἔσω πάντα πρώτου.*

extremity of each thing, i.e. the first thing outside of which no part [of the thing] is to be found, and the first thing inside of which every part [of the thing] is to be found.' Aristotle's definition of 'infinite' in *Physics* III 6 makes the connection clear: 'So, something is infinite, of which it is always possible . . . to take some part outside.'[35] In addition, Aristotle connects being 'whole' and 'complete' with having a limit (i.e. not being infinite);[36] according to Aristotle, anything which is whole and complete and not infinite has a limit. There are no infinite bodies, so all bodies have a limit, called variously *epipedon* or *epiphaneia*,[37] both of which are usually translated 'surface'; it is something which has length and breadth, but no depth (and hence something with the quantity of an area, but not a volume).[38]

Aristotle thinks that every body has an *epipedon* or *epiphaneia*, for the simple reason that all bodies have an outside, where they stop (and where, if there is something outside them, the world outside begins). However, *we* do not readily talk of the *surface* of all types of object: 'it is . . . wrong to imply that everything has a surface. Where and what exactly is the surface of a cat?'[39] Perhaps it is true that the word 'surface' has very precise conditions of application.[40] However, anything which is not infinite (*apeiron*) has a limit (*peras*), and so even a cat has a *peras*, and therefore, since a cat is a sort of body, it has an *epipedon* or *epiphaneia*. The fact is that there is no other word in English except 'surface' for the limit of a body; in using it to refer to the outside or limit of any finite body, I am probably 'ruining the perfectly good word "surface"'.[41]

Do objects have one surface, or several? We regularly use both

[35] 207^{a}7–8: *ἄπειρον μὲν οὖν ἐστιν οὗ . . . αἰεί τι λαμβάνειν ἔστιν ἔξω*. The missing words are *κατὰ τὸ πόσον λαμβάνουσιν*; Aristotle has by this stage of the argument ascertained that to say that something is infinite is to make a predication in the category of *quantity*.

[36] 207^{a}8–9: *οὗ δὲ μηδὲν ἔξω, τοῦτ' ἔστι τέλειον καὶ ὅλον*; 207^{a}14–15: *τέλειον δ' οὐδὲν μὴ ἔχον τέλος· τὸ δὲ τέλος πέρας*.

[37] See e.g. *Phys.* III 5, 204^{b}5–6: a body is *τὸ ἐπιπέδῳ ὡρισμένον*. Aristotle uses *ἐπίπεδον* and *ἐπιφάνεια* interchangeably and sometimes even in the same sentence (*Cat.* 6, 5^{a}1–4; *Phys.* IV 1, 209^{a}7–10); except when the appearance of a body is in question, when he uses *ἐπιφάνεια* exclusively (*Phys.* IV 3, 210^{a}29–30 etc.). But Euclid (*Elements* I, def. 7) defines a plane (*ἐπίπεδον*) as a flat *ἐπιφάνεια* ('surface'), and this difference became rooted in Greek technical vocabulary, prompting Simplicius, for instance, to remark on the fact that Aristotle uses *ἐπίπεδον* for surfaces which are not flat (*In Phys.* 587. 18).

[38] See *Meta.* *Δ* 13, 1020^{a}13–14, with *Cat.* 6, 4^{b}24.

[39] Austin [84], 100.

[40] See Stroll [129].

[41] Austin [84], 100.

the expressions 'the limit' and 'the limits' (cf. 'the surface' and 'the surfaces') of something. Aristotle thinks that in one sense, all non-spherical objects have many surfaces;[42] that a sphere has one surface, but a cube has six. He is quite right to insist on this, for we want to be able to say that something has a flat surface. But if the surface of a solid is taken to be the whole of its outside, there can be no flat surfaces—no solid has an entire outside which is flat, because the outside of an object encloses it completely and hence cannot be flat. But in that case, there are no flat surfaces, and hence no planes, and no geometry.[43] There is a second sense in which something has more than one surface: any object which envelops another—roughly speaking, any object with an internal hole—has its outer surface or surfaces and its inner surface or surfaces.

I have insisted that a cube, for instance, can be seen as having six surfaces, and that if we are to define 'surface' as a limit of a body, and 'plane' as a flat surface, then it is essential that objects can be seen as having many surfaces. However, Aristotle uses a notion of 'limit' in the context of his theory of place according to which we can talk as if these surfaces are collected together as one limit, the idea being that something has just one outside, even if this outside is naturally divided up into different surfaces. In this sense, bodies have, by and large, one surface; and any body which completely contains another body will have two surfaces: one outer one, and one inner one.[44]

When Aristotle talks of the limit at which *x*'s surroundings, *y*, are in contact with *x*, he means to pick out the inner limit of *y*'s two limits.[45] In fact, bodies may well have more than one inner surface—for instance, if there are two swimmers, A and B, in the pool, then there will be two inner surfaces of the water. One way to refer to these different inner surfaces is by referring to the object which they enclose: hence we distinguish between the inner surface of the water at which it is in contact with A, and the inner surface of the water at which it is in contact with B, even though in each

[42] *De caelo* II 4, 286^{b}24–5, talking of a sphere: *μόνη γὰρ περιέχεται μιᾷ ἐπιφανείᾳ* ('[the sphere] alone is enclosed by one surface').

[43] Euclid *Elements* XI, def. 2: *στερεοῦ δὲ πέρας ἐπιφάνεια* ('an extremity of a solid is a surface'); I, def. 7. 'a plane [*ἐπίπεδον*] is a flat *ἐπιφάνεια*. See n. 37 above.

[44] It might even be possible to consider outer and inner surfaces as together constituting one limit of the body.

[45] Cf. 'the *first* [*πρῶτον*] limit of the surrounding body' in his refined definition of a place, at IV 4, 212^{a}20–1.

case the water referred to is the same water. This is why Aristotle refers to 'the limit of the surrounding body ⟨at which it is in contact with that which is surrounded⟩'.[46]

In Chapter 2 I expressed some caution as to whether we could show that the most precise place in which something, *x*, is, would always be a *surface*. But on the view of limits under discussion here, any body, even a complex one, surrounding *x* has an inner limit. If the complex body in question surrounds and is in contact with *x*, then it must come to a stop at that limit.

Now, limits depend ontologically on their hosts (not vice versa): a limit is the limit of something.[47] (Indeed, things have limits because they have such-and-such a form.[48]) But despite this *ontological* inferiority, limits have *explanatory* power. For instance, Aristotle thinks that something is pale because its surface is pale;[49] or that two things touch because their boundaries coincide.[50] The inner limit of *x*'s surroundings depends on those surroundings for its existence, but it has explanatory power. For *x* is in contact with its surroundings in virtue of the fact that *x*'s outer boundary is coincident with the inner limit of its surroundings.

It is natural to assume that the limit of *x* marks *x*'s end, and the inner limit of *x*'s surroundings marks the beginning of the outside world. The inner limit of what surrounds me is, if you like, the first contact I have with the outside world. Moreover, the inner limit of what surrounds *x* is that in which *x* is non-derivatively; it is the first thing in which *x* is. It explains why *x* is in any of the things in which it is. It is in virtue of being in it that *x* is in anything else.

The individuation of limits

Because the inner limit of *x*'s surroundings is that in virtue of which *x* is in anything at all, Aristotle thinks of it as being *x*'s proper place. But there is more to say. Consider the sentence: '*x* is near *y*'. This is

[46] 212^{a}6–6*a*. See n. 2 above.

[47] See e.g. *Meta.* Z 10, 1034^{b}28–32, with Barnes [88], 293 n. 128.

[48] The bronze (i.e. the matter) of a bronze sphere has no surface; it is the sphere (i.e. the matter with the form) which has a surface. A form gives shape to its matter: the form defines or limits the matter. See e.g. *Phys.* IV 2, 209^{b}3–4.

[49] See e.g. *Phys.* IV 3, 210^{a}29–30: λευκὸς ὅτι ἡ ἐπιφάνεια λευκή.

[50] See *Phys.* IV 4, 211^{a}33–4 ; V 3, 226^{b}23. Cf. Chapter 4, p. 130.

a local or spatial predication. Aristotle does not appear to think that it says where *x* (or *y*) is—it is not a predication in the category of where—but it certainly has something to do with places. It seems that we can paraphrase it as '*x*'s place is near *y*'s place'. (This formulation brings out the fact that if *a* were where *x* is, and *b* were where *y* is, then *a* would be near *b*.) Following Aristotle's analysis, '*x*'s place is near *y*'s place' analyses as 'the limit of that which surrounds *x* is near the limit of that which surrounds *y*'. If *a* were surrounded by the same limit as *x*, and if *b* were surrounded by the same limit as *y*, then *a* would be near *b*.

Talking of limits in the context of a predication such as '*x* is near *y*' is welcome. For at the very least, *x* is near *y* iff the (or perhaps an) edge of *x* is near the (or perhaps an) edge of *y*. (This is not true for other parts of *x* and *y*—*x* could be near *y*, but their midpoints, or centres of gravity, could be far apart. France is near Britain, but Nice is not near Edinburgh.) None the less, consider the inner boundary of *x*'s surroundings. It is coincident with *x*'s boundary. If something else, *z*, were surrounded by that boundary, then it would be near *y*. That is, it is not *x*'s boundary which determines whether *x* is near *y*, but rather the inner boundary of *x*'s surroundings; if it were to surround something else, *z*, then *z* would also be near *y* (assuming *y* did not move).

For this to work, we need to attach a sense to the idea that two different things can be surrounded at different times by the same limit. This is essential also for even simpler cases. 'Aristotle's account of place in the *Physics* rests squarely on the assumption that A can move into the identical place vacated by B.'[51] We want, for instance, to be able to say '*x* is now in the same place as *y* was'. But if we analyse this as '*x* is now surrounded by the same limit as *y* was', then we use the notion of 'the same limit', and we need some conditions of individuation. Conversely, if the limit surrounding *x* at t_1 is different from the limit surrounding *x* at t_2, then *x* will have moved. If such conditions of individuation can be provided, then it looks as if the inner limit of *x*'s surroundings is an extremely strong candidate for being the place of *x*.

The first step in giving these conditions is made explicitly by Aristotle himself. At *Metaphysics B* 5, 1002^{a}4–14, he wonders whether a surface, a line, and a point are not, after all, substances.

[51] Owen [124], 102.

Owen[52] underlines the fact that this is at least in part because the boundaries of something are not like, say, the shape of something; it seems that two things can have the same shape, but not the same boundaries. So if *x* and *y* are distinct, then *x* and *y* have different limits. The next step is to observe that there is a connection between the boundaries of something on the one hand, and its shape on the other. This was a familiar Platonic observation.[53] When something changes shape, it changes boundaries. But also, when something changes size, it changes boundaries. Thus, the outer boundary (surface) of something is individuated by it, its size and shape.

Here are some corollaries. A bronze sphere which is rolling around on the table does not change boundaries. However, if it shrinks (perhaps bronze shrinks at certain temperatures), while remaining spherical, it does change boundaries: before, its surface was of a certain surface area, for instance, but after shrinking it has a different surface area. Similarly, if the sphere grows (perhaps bronze expands at certain temperatures), then its surface changes. (Two surfaces are different if they are not of the same surface area.) Finally, when something changes shape, then its boundaries change. Thus, the sea is changing in respect of its surface (undulating), as the waves come in.

Given this rough account, surfaces are assumed and discarded just like shapes, sizes, and colours. Something keeps the same surface as long as it has the same shape and size, and it regains that surface if ever it has again the same shape and size. Once again, this does not seem right as far as one particular use of the word 'surface' goes. For instance, if I take the top of this table off, and replace it with the top of another table, then my original table may well have the same shape and size as before, but it does not have the same surface. (It might be possible to object in this instance that it is no longer the same table, but there are certainly formulations of the same problem where this objection would not hold.) But, according to the conditions which I have given, the table has the same boundaries as before.

We have put the burden of individuation on the notions of a body,

[52] Ibid.: 'plainly my desk cannot be said to have the same (as against the same sort of) surface, or edge, or corners, as any other'.

[53] 'Figure [σχῆμα] is the limit [πέρας] of a solid' (*Meno* 76 A). Aristotle notes this at *Meta.* Δ 17, 1022ᵃ5–6: '[a limit] is also the form [εἶδος] of a magnitude or that which has magnitude'. For the identification see Chapter 4, p. 108.

size, and shape. But perhaps we shall find that where we had difficulties with the concept of a surface of something, we have problems with its shape. And indeed, Austin challenges: 'What is the real shape of a cat?'[54] But this is a different challenge from the one he threw down about surfaces. For he points out that it is difficult to attribute one—'real'—shape to a cat, because it changes shape all the time: it gets thinner and fatter; it gets up, runs, stretches, curls up. It changes limit: it always has a limit (it does not become infinite during these changes) but it takes on and discards different limits throughout all these changes.[55]

What I have said so far has dealt with the outer boundary of something. But Aristotle's account of places deals with the *inner* limit of something, i.e. with the inner boundary of something when it contains something else. Presumably, we have to talk of the external and internal shape of something. External shape is a question of how the object is in contact with what is outside it; for instance, a squash ball and a golf ball could have the same external shape, even if a squash ball is hollow, and a golf ball is not. Internal shape is to do with how the object is arranged internally.

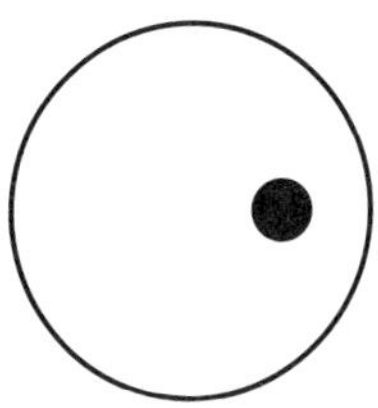

FIGURE 1

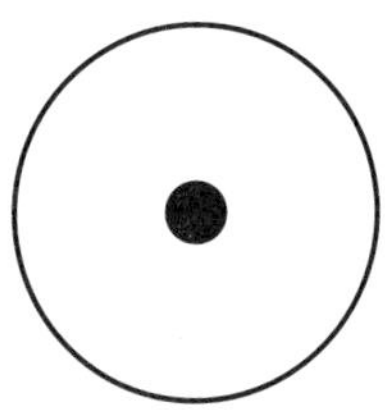

FIGURE 2

Take the discs in Figures 1 and 2, each with a hole. The two figures represent objects with different shapes; for instance, one is a good shape for an LP musical disc, the other is not. Their external shape is the same, they differ in shape internally. This is not (just) a question of the shape and size of their holes, for in our example they are the same shape and size. Rather, it is a question of where

[54] Austin [84], 67.

[55] There is a subtle objection to this rough account of the individuation of limits or boundaries. Suppose France were redrawn with the same shape and size, but ten miles to the west. (The *land mass* does not move: colonial powers simply redraw the boundaries of France.) France has changed boundaries, frontiers—but it is exactly the same shape and size as before. Perhaps the only way to answer this is to say that France is a rather peculiar entity (no colonial power can redraw the boundaries of *substances* or *bodies*).

the hole is within the disc, or rather, how the objects are internally disposed.

So suppose there is an amoeba moving around in a dish of water. It starts off towards the right (Fig. 1), and then moves to the centre (Fig. 2). The amoeba is in contact with the water at a *different* inner limit in Fig. 1 from the inner limit in Fig. 2, because the internal shape of the water has changed.

Various puzzles arise when we consider the problem that something may contain more than one thing. Suppose there are two fish in the fish tank. If one fish is immobile and the other in motion, then we have to say that the water is changing shape. But one inner limit stays the same; the other changes. We need to say that the water is changing shape 'around' the moving fish, but not around the immobile one. I shall not attempt to produce a detailed account here. I think it is obvious that we can make use of the idea that the water changes shape around one fish, but not the other, without adverting to places; dispositions of parts, shapes, and sizes are concepts which are independent of the problems of place. Hence Aristotle can use the notion of an inner limit (which I claim is connected to the notion of internal shape) to elucidate the notion of *place*.

Places

Aristotle's first definition of the proper place of x runs as follows: the limit of the body which surrounds x, at which it is in contact with x. 'The body which surrounds x' refers to x's surroundings, and I have argued that this should be understood as being x's maximal surrounder. This indeed yields the right results. For according to the conditions of individuation for a limit, the following is true:

(1) The inner limit of the maximal surrounder of x will change iff x changes place within that maximal surrounder.

An analogy should help bring out the importance of (1). Suppose there is a fish in a tank of water. As the fish moves around in the tank, the water changes form around the fish—it changes 'internal shape' around the fish—and hence is in contact with the fish at a different boundary. But if the fish does not move, the water remains in the same shape around the fish—at the same boundary. This is just an example of *antiperistasis* ('mutual replacement').[56] The insight to hang on to here is that something which is immobile and which surrounds *x* during a period in which *x* moves must change internal shape around *x* to accommodate *x* as *x* changes place. In other words, its parts redispose themselves around *x*. As Aristotle says: 'it is possible for things to make way for one another simultaneously'.[57]

In this analogy, the water is analogous to the universe and the fish to an occupant of the universe. Just as the contents of the tank (as a whole) cannot move sideways, the universe cannot move in any direction. For the universe is full and cannot move. How could it move? There is nowhere for it to go—there is nothing outside of it.[58] There is nothing with which it can exchange places. Thus, every movement of something in the universe involves the rest of the universe rearranging itself, just as any movement of the fish involves the water rearranging itself around the fish.

One way in which Aristotle talks of the limit between *x* and the rest of the universe is as follows: 'The place [of a changeable body] is not the universe, but some [part] of the universe, which is a boundary and in contact with the changeable body.'[59] Something's proper place is the limit where the universe is in contact with it. As far as the analogy with the fish in the tank of water goes, this would be expressed as 'the place of the fish is the limit of the contents of the tank where they are in contact with the fish'. (If we define *x*'s maximal surrounder as 'everything except *x*', then nothing other than *x* could ever be surrounded by it, and nothing could ever take one of *x*'s places. But it is clearly central to Aristotle's conception of place that one thing can take something else's place, for he introduced places with the example of replacement—when the air takes the place of water, as in a vessel.[60] The maximal surrounder of *x* should be understood as the whole universe—the common place of all things.) *x* is in the universe, and there is a

[56] See Chapter 1, pp. 23–5, and Chapter 4, pp. 120–1.

[57] *Phys.* IV 7, 214^{a}29–30: ἅμα γὰρ ἐνδέχεται ὑπεξιέναι ἀλλήλοις.

[58] *Phys.* IV 5, 212^{b}9–10: μηδὲν αὐτὸν περιέχει σῶμα. Something outside of which there is nothing can rotate, but it cannot go somewhere else. See 212^{a}31–^{b}3, and 'Rotation and the place of the universe', pp. 166–9 below.

[59] *Phys.* IV 5, 212^{b}18–19: ἔστι δ' ὁ τόπος οὐχ ὁ οὐρανός ἀλλὰ τοῦ οὐρανοῦ τι τὸ ἔσχατον καὶ ἁπτόμενον τοῦ κινητοῦ σώματος. Aristotle means *proper* place; the universe is the *common* place of everything.

[60] *Phys.* IV 1, 208^{b}1–3: 'That place exists seems clear from replacement: for where there is now water, there, when it goes out as from a vessel, air in turn is.'

limit where the universe is in contact with x. The inner limit of x's maximal surrounder just is the inner limit of x's surroundings. Aristotle's idea is that if the universe assumes the same internal shape around y as it had assumed around x, then the inner limit of the universe—the inner limit of x's surroundings and the inner limit of y's surroundings—will be the same in each case.

Elucidated in this way, (1) can be seen to be true, and we can add (2):

(2) The inner limit of the universe at which it is in contact with x at t_1 is the same as the inner limit of the universe at which it is in contact with y at t_2 iff y is in the same place at t_2 as x was at t_1.

One corollary is that only things with exactly the same shape can have exactly the same proper place as something else. However, this is bound to be the case given the condition which Aristotle puts on what a proper place is, namely that it 'hug' its occupant, and be the same size as it.[61] This conception of place is demanded by principles such as 'two bodies cannot be in the same place at the same time'. But Aristotle's account allows for the places of two things to be specified in more *general* terms in such a way that it is possible to say of two items that they are both in such-and-such a body, e.g. that they are both in the air. Of course, we know that two distinct items will not be in exactly the same place, but (put in another way) they will always share some broad places—often when we ask where two things are, we wish to be told some broad place which includes them both.

The immobility of places

Something which has a place has surroundings. In a broad sense, something's surroundings are the universe. The universe is immobile, for there is nothing relative to which it can move, nothing which it can displace. Aristotle identifies the proper place of x with

[61] This is one of the *ἀξιώματα* at the start of IV 4: [*ἀξιοῦμεν*] *τὸν πρῶτον* [*τόπον*] *μήτ' ἐλάττω μήτε μείζω* (211ᵃ2). (Cf. IV 1, 209ᵃ28–9.) For a use of the *ἀξίωμα*, see IV 4, 211ᵃ27–9: *εἰ γὰρ πᾶς ὁ ἀὴρ τόπος, οὐκ ἂν ἴσος εἴη ἑκάστου ὁ τόπος καὶ ἕκαστον, δοκεῖ δέ γε ἴσος εἶναι, τοιοῦτος δ' ὁ πρῶτος ἐν ᾧ ἐστιν* ('If the whole air were its place, the place of each thing and each thing would not be equal, but it seems that it is equal; such is the primary place in which each thing is').

the inner limit of the universe at which the universe is in contact with *x*, and so something changes position in the universe iff it changes proper place. Whenever something moves from somewhere to somewhere else, the universe as a whole indulges in *antiperistasis*—reorganization within a *plenum*. The universe rearranges itself in response to moving items, such that we can talk of the form of the universe at any one time around these moving items. This enables us to identify the proper place of *x* as the inner limit of the universe at which it is in contact with *x*. The definition of a place at 212^{a}6–6*a* reads: 'the limit of the surrounding body ⟨at which it is in contact with that which is surrounded⟩'—'the surrounding body' should be understood as referring to the maximal surrounder.

The boundary of the outside world at which it is in contact with something, *x*, is an item, not a relation, or something abstract like that. (Aristotle's concept of place is perhaps relational, but it does not identify places with a relation.[62]) It is the most precise place where *x* is, that is, it is the most precise thing in which *x* is. Suppose I walk from one end of Athens to the other between t_1 and t_2. Then, broadly speaking, I am in the same place at t_1 and t_2—I am in Athens both at t_1 and at t_2. But there is somewhere where I am at t_2 where I was not at t_1. An account of place which wants to do justice to this must identify something's proper place as something which will always change when something moves from being somewhere to being somewhere else (I emphasize that this leaves open a broader conception of place which can also account for the idea that I can move and still remain in the same broad place—for instance, in the example given, I remain in Athens, even though I have moved).

It is obvious, therefore, that the proper place of something cannot itself move. Otherwise, something could move from being somewhere to being somewhere else, without changing place. Aristotle identifies the proper place of *x* at time *t* as the inner limit of the universe at which it is in contact with *x* at *t*. Because the universe cannot move, there is no danger that something which moves from being somewhere to somewhere else could ever be surrounded by the universe at the same boundary.

In this sense, things have places relative to the immobile universe. The universe is the common place of all things and so its immobility guarantees that the frontier between it and its occupants can count as their proper place, and its immobility is what lies behind the fact

[62] See below, pp. 169–71, for more on the relative–absolute distinction.

that something which moves displaces the contents of the universe: 'the change of a moving body occurs in its surroundings, which are at rest'.[63] The fact that the universe is immobile means that it will rearrange itself around something which moves from somewhere to somewhere else.

Consider a moving body, x. x moves through the universe—the inner boundary at which the universe is in contact with x changes as x moves from one proper place to another. Sometimes, however, there will be smaller items through which x moves—more precise places, to use the earlier terminology. Places have to be immobile, so Aristotle insists that only those immobile intermediate items count as places.[64] Here is what he says:[65]

> Just as a vessel is a movable place, so is a place an immobile vessel. For this reason, whenever that which is inside moves and changes in something which is itself moving, for instance a boat on a river, it uses its surroundings as a vessel rather than as a place. But its place wants to be immobile. For this reason, the whole river is a place, because the whole river is immobile. So, the first immobile limit of that which surrounds, that is the place.

Aristotle imagines that one of the things in which the boat is at t_1 is actually in motion at time t_1 (the boat 'moves and changes in something which is itself moving'). He is thinking of a certain mass or body of river water, q, which surrounds the boat at t_1. As the boat travels downstream, q is also moving,[66] and could in theory travel with the boat, such that at a later moment, t_2, the boat is still surrounded by q, at exactly the same inner boundary of q as at t_1. If, therefore, we take q to be the boat's relevant surroundings, the inner limit of the boat's surroundings at t_1 and t_2 will be the same.

[63] IV 4, 212ᵃ9–10: *ἐν ἠρεμοῦντι τῷ περιέχοντι γίγνεσθαι τὴν μετάστασιν τοῦ φερομένου.* Aristotle adduces this as a reason for why places are a difficult subject: people have been led by this fact into thinking that places are the gap between the limits of the (immobile) surroundings.

[64] Hence the definition given in Chapter 2 above (that a place of x is anything in which x is) is not quite refined enough.

[65] IV 4, 212ᵃ14–21: *ἔστι δ' ὥσπερ τὸ ἀγγεῖον τόπος μεταφορητός, οὕτως καὶ ὁ τόπος ἀγγεῖον ἀμετακίνητον. διὸ ὅταν μὲν ἐν κινουμένῳ κινῆται καὶ μεταβάλλῃ τὸ ἐντός, οἷον ἐν ποταμῷ πλοῖον, ὡς ἀγγείῳ χρῆται μᾶλλον ἢ τόπῳ τῷ περιέχοντι. βούλεται δ' ἀκίνητος εἶναι ὁ τόπος· διὸ ὁ πᾶς μᾶλλον ποταμὸς τόπος, ὅτι ἀκίνητος ὁ πᾶς. ὥστε τὸ τοῦ περιέχοντος πέρας ἀκίνητον πρῶτον, τοῦτ' ἔστιν ὁ τόπος.*

[66] See *Meteor.* II 2, 354ᵇ13–14: *τὸ γὰρ τῶν ποταμῶν οὔτ' ἀθρόον οὔτε στάσιμον, ἀλλ' ὡς γιγνόμενον ἀεὶ φαίνεται καθ' ἡμέραν* ('River water is not a unity [*ἀθρόον*], nor is it stationary [*στάσιμον*], but is seen to be in a continuous process of becoming from day to day').

So, if one identifies the place of the boat at t_1 as the inner limit at t_1 of q, and its place at t_2 as the inner limit at t_2 of q, then the boat will be deemed to be in the same place. However, the boat has moved. Thus q does not have the property that the universe has, namely, that it surrounds the boat at a different limit iff the boat moves. The mass of river water in question, q, acts as a vessel which has moved, carrying its contents (the boat) with it, rather than as a place of the boat. It has acted as a vessel in so far as it has moved *with* its contents (the boat) and surrounds its contents at the same boundary, just as when I pass you a pitcher of water; water and pitcher move together, and the pitcher surrounds the water at the same boundary or surface.

However, between t_1 and t_2, the *whole* river does not move,[67] unlike the mass of river water we were considering, q. (The Thames does not move: it remains where it is, even if the water which constitutes it at any one time is in full flow.[68]) Because the whole river does not move, it is true that, while the boat is on the river, the inner limit of the river at which it is in contact with the boat will change iff the boat moves. Hence, while the boat is on the river, we can say where the boat is by locating it relative to the river, for instance by saying in which part of the river the boat is ('the stretch between Oxford and Abingdon'). The boundary at which the river is in contact with the boat is then the most precise part of the river on which the boat is.

Aristotle seems to imply that the inner limit of the whole river where it is in contact with the boat is different from the inner limit of the water where it is in contact with the boat, for when he says that the proper place of the boat is the first *immobile* limit, he apparently implies that there are mobile limits and immobile limits at which something is in contact with its surroundings. But the expression 'immobile limit' is misleading. Limits are things which are *assumed* and *discarded*, like shapes, sizes, colours, etc. They are not the sort of thing which can move; when their host body moves, they move *derivatively*.[69] Limits *inherit* mobility or

[67] 212^{a}19–20: *ἀκίνητος ὁ πᾶς* [sc. *ποταμός*]. I have much more to say about *ἀκίνητος* below, pp. 155–61.

[68] This way of understanding Aristotle is due to Burnyeat [43], 232 n. 15. He points out that 'the river . . . as a geographical entity' is 'static'. Much depends on what 'static' means (see n. 67 above).

[69] Cf. *Phys.* IV 4, 211^{a}21–2: *τὰ δ' οὐκ ἐνδεχόμενα* [*κινεῖσθαι καθ' αὑτό*] *ἀλλ' αἰεὶ κατὰ συμβεβηκός, οἷον ἡ λευκότης καὶ ἐπιστήμη* ('other things cannot [move in their

immobility from their hosts. Thus, the reference to immobility in 'the first immobile limit of that which surrounds' applies (strictly speaking) to a host of the limit.[70]

However, there *is* only one boundary in question; that between the boat and its surroundings, however much of those surroundings are in question. According to how much, or how, we pick out the boat's surroundings, we pick this boundary out under different descriptions. (Here, there is no question of a different limit, even if we are picking out different bodies which are host to the limit, because the boundary is actually common to all the bodies.) If we pick out the inner limit of x's surroundings at t_1 as the inner limit of something in motion, this will not give us the result that x is in the same (proper) place as t_1 at t_2 iff x is surrounded by the same limit at t_2 as at t_1.

Aristotle's example is deliberately chosen in order to recall a well-known problem in ancient natural philosophy. No *phusikos* ('student of nature') will have been unaware of Heraclitus' famous puzzle that you cannot step into the same river twice,[71] one of the formulations of which is uncannily suggestive of our problem: 'On those who step into the same rivers, different and different waters flow.'[72] What is true of the river *qua* river is not necessarily true of the river *qua* body of water, even though at any one moment the river is not distinct from that body of water.

Adopting this terminology, we might say that at t_1 the inner limit of the boat's surroundings *qua* inner limit of the river water is not the boat's proper place; the inner limit of the boat's surroundings *qua* inner limit of the river is the boat's proper place. This is because *qua* limit of the river, it is immobile; *qua* limit of the river water, it is moving (albeit derivatively). Limits derive properties (e.g. movements) from their hosts, so we have to distinguish dif-

own right] but always move derivatively [κατὰ συμβεβηκός], such as pallor and knowledge').

[70] Cf. Burnyeat [43], 232 n. 15: '"the innermost *static* boundary of the surrounding body" . . . is equivalent (boundaries being what they are) to "the innermost boundary of the surrounding static body"' (his italics).

[71] 22 B 91 DK. See Barnes [27], 66, for discussion.

[72] 22 B 12 DK. The formulation is close to our problem because if we attempt to give the place of a *stationary* boat on a river by adverting to the *water* which surrounds it at any one time, we can see that the boat will be deemed to change place; similarly, my foot is surrounded by different water each time I step into the river, even if each time I step in at the same place.

ferent ways of picking out the same limit, according to its hosts. At t_1 there are many different things in which the boat is. Some are in motion. But only those which are immobile are being used as places by, or serve as places for, the boat.[73]

The passage I cited above from *Physics* IV 4 contains the kernel of Aristotle's concept of place. Although at t_1 there are many different bodies which surround the boat, only some of them will serve *over time* as a place of the boat. They serve as a place in so far as they are immobile. *Just as the immobility of the universe guarantees that we can refer* x *to the universe in order to say where it is, so the immobility of any immobile intermediate surrounders of* x *guarantees that we can refer* x *to them in order to say where* x *is.* For particular journeys which *x* undergoes, the rearrangement of the universe around *x* can sometimes be isolated as occurring in *x*'s more immediate surroundings. In our example the whole river is just such a body. The boat uses the river as a place while it is on the river, because it will be in virtue of any change in relation of the boat to the river that the universe as a whole indulges in *antiperistasis*.

I have made use of the notion of 'picking something out under a description', and of *qua*-operators. This is to disguise the following awkward fact: there is only one inner boundary in question. But surely either this boundary is the proper place of the boat, or it is not. The means we use to refer to it should not make a difference. The inner boundary of the universe at the time in question just is the inner boundary of *q*, and just is the inner boundary of the river. If boundaries inherit properties from their hosts, then the boundary in question has contradictory properties—notably, being the proper place of the boat and not being the proper place of the boat.

There is, I think, no cause for alarm. When I say that the inner boundary of the body of water *q* is nothing other than the inner boundary of the whole river, it is not clear that 'is nothing other than' should really be taken as being an expression of *identity*. Rather 'is nothing other than' (or 'is simply' or 'is merely' or 'just is', even just 'is') should be taken as an expression of composition or constitution, not identity. This use of 'is' is not an *ad hoc* inven-

[73] IV 4, 212^{a}17–18: The boat on the river, when the river is considered as a quantity of river water in motion, 'uses its surroundings as a vessel rather than a place' (ὡς ἀγγείῳ χρῆται μᾶλλον ἢ τόπῳ τῷ περιέχοντι).

tion but is the subject of much philosophical scrutiny and is often invoked to overcome paradox in discussions of the composition of spatio-temporal continuants such as humans, who are composed at various times of differing collections of matter and yet appear to be nothing more than this matter.[74] The view which emerges, then, is that the limit at which the universe is in contact with a stationary object remains the same, even though it is composed at different times by the limits of all sorts of entity. (Remember also that in most cases it is composed of the limit of more than one body at any one time; for instance, I am now in contact with both air and ground.) So a stationary fish in water is surrounded by many ever-changing surfaces of water, which constitute the inner surface of the sea at which it is in contact with the fish; this surface in turn constitutes the inner surface of the universe at which it is in contact with the fish. The inner surface of the universe at which it is in contact with the fish (as long as the fish remains stationary) remains the same, even though it is constituted at different times by different surfaces of water. Suppose the fish were to remain stationary (sitting at the bottom of a rock pool, perhaps), and the water around it were to drain away so that it is left exposed to the air. Then the surface of air around the fish would constitute the inner limit of the universe at which it is in contact with the fish. That limit would remain the same (because it is individuated by the internal shape of the universe around the fish), yet it is (or is composed of) the inner limit of air then, and was (or had been composed of) the inner limit of the sea before.

The standard objections to Aristotle's definition of place

Before getting into the difficult territory of what exactly *akinētos* means, I want to be quite clear about how Aristotle's definition of place stands up to two standard objections made to it. The two objections run as follows:[75]

(1) When there is a stationary boat on a river, the surrounding body (the water) is in contact with the boat at constantly *different* limits, because the water is changing all the time. Hence the proper place of the boat cannot be this limit.

(2) When there is a moving boat on a river, the surrounding

[74] See e.g. Wiggins [132], 30 ff.

[75] See e.g. Sorabji [26], 187–92; Algra [21], 222–30.

> body (the water) could be in contact with the boat at the *same* limit, because the boat moves with the water. Hence the proper place of the boat cannot be this limit.

Objection (2) is considered explicitly by Aristotle in the passage about the river. But objection (1) is countered in exactly the same way. The answer is this. Do not suppose that the surrounding body must be the water of the river. The surrounding body should be taken, *in the first instance*, to be the maximal surrounder of the boat. In case (1), its inner limit around the boat *does* remain the same during the time in question, and in case (2), its inner limit *does* change during the time in question (the 'internal shape' or arrangement of the universe changes around the boat). But Aristotle thinks that there is an *intermediate* surrounder, namely the whole river, whose inner limit *also* remains the same around the boat in case (1), and changes around the boat in case (2), namely the whole river. Thus, the whole river *also* qualifies as a place of the boat. The inner limit of the whole river (since the whole river is immobile) can be *identified* with that of the maximal surrounder of the boat. The inner limit of the water of the river cannot, because the water (only) *composes* the river. The inner limit of the whole river is the first immobile limit of the boat's surroundings. The inner limit of the water of the river is not.

I have tried to show how we have a choice when we specify something's surroundings of how precise or imprecise we are (this is entirely parallel to being precise or imprecise about something's place). Aristotle puts conditions on *which* intermediate specifications of something's surroundings qualify as specifications of that thing's *place*. That qualification is that the intermediate surrounder should be immobile—because only then will its inner limit change around the thing as that thing moves.

The meaning of *akinētos* ('immobile')

The intermediate places between the common place of all things (the universe) and the proper place of each thing (the boundary at which the universe is in contact with each thing) will never be immobile in quite the sense that the universe is immobile, or the sense in which the inner boundary of the universe is immobile. This is because these intermediate places will be bodies which have surroundings, and hence which could, in theory, move. The river as a whole is one example; if it is immobile, it is so only by accident, for rivers can be displaced. (Two ancient natural philosophers were known as river-movers, namely Thales[76] and Empedocles.[77]) Aristotle seems to indicate that the river is not immobile in *quite* the required way when he says that the boat *uses* the river *as* a place.

In fact, *akinētos* has many meanings. For a start, it could have either middle or passive meaning. For instance, the 'Prime Mover' is *akinētos*, in the sense that nothing moves it. But *akinētos* as said of places and the river is middle in sense, meaning something like 'not moving' (i.e. 'not in motion'). Furthermore, verbal adjectives ending '-*tos*' have many nuances, and Aristotle himself distinguishes some of such nuances in the meaning of *akinētos* towards the end of *Physics* V 2, 226^{b}10–15:[78]

> The term *akinētos* we apply (1) to that which is absolutely incapable of being moved (just as we correspondingly apply the term 'invisible' to sound); (2) to that which is moved with difficulty after a long time or whose movement is slow at the start—in fact, what we describe as hard to move; and (3) to that which is naturally designed for and capable of motion, but is not in motion when, where, and as it naturally would be so.

(*Kinēsis* here has been translated by cognates of 'move', which appears to be the sense in question in *Physics* IV 4 in so far as Aristotle says that the whole river is *akinētos*, yet rivers can and do constantly change in other senses: their water level rises and falls, they ice over, etc.)

There are middle senses of *akinētos* too. As Aristotle says, oysters cannot move, they are *akinētos*:[79] this does not mean that they are *difficult to* move (luckily, or else it would be difficult to get them to the dinner table), nor that they have difficulty getting going, but rather that they do not *naturally* move, or do not have the *faculty* of movement. One other sense which is important for us is the sense in which the universe is *akinētos*. It is a body, and hence is the sort of thing which *could* move or be moved, and so is not *akinētos* in sense (1). On the other hand, there is nowhere for it to go, and so

[76] 11 A 1, § 38, and A 6 DK.

[77] 31 A 1, § 70 DK.

[78] *ἀκίνητον δ' ἐστὶ τό τε ὅλως ἀδύνατον κινηθῆναι, ὥσπερ ὁ ψόφος ἀόρατος, καὶ τὸ ἐν πολλῷ χρόνῳ μόλις κινούμενον ἢ τὸ βραδέως ἀρχόμενον, ὃ λέγεται δυσκίνητον, καὶ τὸ πεφυκὸς μὲν κινεῖσθαι καὶ δυνάμενον, μὴ κινούμενον δὲ τότε ὅτε πέφυκε καὶ οὗ καὶ ὥς.* Translation adapted from the Oxford translation [1]. The numbering is mine.

[79] See *HA* I 1, 487^{b}14–15.

it is not exactly *akinētos* in sense (3); it is not that it just *happens* not to be moving. Nor is it *akinētos* in the way an oyster is (i.e. not possessing the faculty of movement) or the Eiffel tower is (difficult to move). The sense in which the universe is *akinētos* I shall call sense (4). As Aristotle himself says in explanation of why the universe is at rest: 'it might be that it had, not moving, nowhere else to go, yet there was nothing to prevent its being by nature such [as to move]'.[80]

Limits are immobile in sense (1). However, they inherit mobility etc. from their hosts. The immobility—sense (4)—of the universe guarantees the derivative immobility—sense (4)—of the proper place of the boat, where this is understood as the boundary at which the universe is in contact with the boat. If there are any intermediate surrounders, they will never be immobile in senses (1) or (4), but they may well be immobile in senses (2) or (3). Hence, these places are not immobile in quite the way that the common place of all things (the universe) and proper places of things are immobile. But they may well for all that not be moving, and *while they are not moving*, they *serve as* places for their occupants in so far as any movement of any of their occupants will entail a change of boundary between them and that occupant. Intermediate places thus differ from proper places and the common place in that they are immobile in a different way.[81]

However, Aristotle argued as follows: (*a*) places want to be immobile (212^{a}18–19); (*b*) so the whole river is a place because *it* is immobile (212^{a}19–20); and so (*c*) the (proper) place is the inner *immobile* limit of what surrounds (212^{a}20–1). Each time, Aristotle uses the word *akinētos* for 'immobile'. He is thus sliding somewhat between senses of *akinētos* each time. Such sliding is perhaps inappropriate, but not, I judge, disastrous. Given that the proper place of something, *x*, is the inner limit of the universe where it is in contact with *x*, the proper place of *x* is guaranteed to be immobile in the strong sense (4). We can say where *x* is by relating it to the universe, because the universe is known to be not moving, for the simple reason that it has nowhere to go. It is then sometimes adequate to refer to this limit as the inner limit of smaller surrounders

[80] *Phys.* III 5, 205^{b}9–10: *εἴη γὰρ ἂν καὶ ὅτι οὐκ ἔχει ἀλλαχῇ κινεῖσθαι οὐ κινούμενον, ἀλλὰ πεφυκέναι οὐδὲν κωλύει.* Aristotle is criticizing Anaxagoras for not having explained *why* the universe is at rest; is it *naturally* at rest, or what?

[81] This is perhaps why Aristotle does not, in IV 2, use 'place' for these intermediate surrounders. See Chapter 2, p. 59.

of x, as long as these are immobile. They can never be immobile in sense (1) or (4), but they may be immobile in sense (3) and even in sense (2) as well. While they are not moving, and for as long as they surround x, they are used as places by x.

To put this in another way: the fact that the inner limit of the universe where it is in contact with x is the proper place of x guarantees that every body (i.e. everything with a limit) in the universe has a place.[82] But this limit can sometimes be considered as that of a smaller body than the universe, and this is particularly obvious in the case where the displacement caused by the movement of something happens within a fixed body such as a river. For in these cases, it is even true that the *antiperistasis* ('mutual replacement') that has to happen when something moves occurs in that intermediate body. Hence, the river is an excellent case of an intermediate item between the universe and the boat, relative to which we can say where the boat is.

For all that, intermediate immobile surrounders need to be approached with care. Let us say that if x uses y as a place, y locates x. Then the river could be termed an 'ephemeral locator', since it will not necessarily locate the boat all the time. (The boat may leave the river; or the river may start to change course.) For any period of time t_1–t_2, if we ask whether x at t_2 is in a different place from the one it was in at t_1, then we have to find something which locates x during the period t_1–t_2. Something which locates x at t_1, and again at t_2, will not necessarily do. (Perhaps it was immobile at t_1 and t_2, but moved during some small period in between, taking x with it. In this case, it might not have a different internal shape around x at t_2 from the one it had at t_1, yet x has changed place.) If ever, over a period of time, x stays in the same place, it is because there is something which locates x during this time, and which has constantly been in contact with x at the same inner limit.

Even this is not ideal. Something, z, may well locate x over a period t_1–t_2, and then again over a period t_3–t_4. However, if some time between t_2 and t_3 z moves, then something could be surrounded by a particular limit of z between t_1 and t_2, and exactly the same limit between t_3 and t_4, and yet be in a different place.

[82] Except—presumably—for the universe itself, and the outer sphere of the universe. Aristotle appears to think that all parts of the universe have a place, because they all surround one another (IV 5, 212^{b}13). But there is nothing surrounding the outer sphere of the universe. We should probably imagine it divided into segments. See 'Rotation and the place of the universe', pp. 166–9.

Ultimately, the universe is what guarantees whether something has moved or not.

Aristotle's account favours specifying some smaller (and possibly ephemeral) locator as the host of the place of x, because smaller locators show how the universe changes around x, and are evidently more in line with our practice of saying where things are (we advert, normally, to intermediate items in something's surroundings—although this depends greatly on context, of course). Aristotle talks of something's primary place;[83] the suggestion behind 'primary' is surely that there are secondary places and so on. He has already referred to the universe as the 'common place', in which all bodies are[84]—everything (except the universe itself) is in the universe. He strongly hinted that both the air and the earth are my broader places, even if they are not my proper place.[85] *Any* adequate interpretation of Aristotle must account for the idea that there are places other than the common place and proper places.[86] Intermediate locators are these other places.

Aristotle explicitly identifies the river as a broad place of the boat. 'Rather, the whole river is a place, since the whole is immobile.'[87] We should not, of course, be tempted to translate this sentence 'the whole river is the place, since it is immobile'. The view that the whole river is the boat's proper place would be utterly unacceptable to Aristotle, for several reasons, not the least of which is that he insists that something's place is the same size as its occupant; moreover, from the fact that the whole river is immobile he immediately draws the conclusion that the place of something is the

[83] At 210^{b}24 and 211^{a}2. This place is the same as the 'particular' (ἴδιος) place of IV 2, 209^{a}33, which is the *first* thing in which something is (ἐν ᾧ πρώτῳ).

[84] IV 2, 209^{a}32.

[85] Cf. the examples of predications in the category of where in the *Categories*, viz. 'x is in the Lyceum' and 'x is in the agora' (*Cat.* 4, 2^{a}1–2). The Lyceum and the agora are clearly typically 'broad' places in this sense: by and large they contain many different bodies, without it being the case that these bodies are coincident. Notice, however, that we are dealing here with receptive containing and not circumscriptive containing, and so it is not clear what 'the inner boundary' of the agora and Lyceum actually would be. See Chapter 2, pp. 56–8.

[86] The distinction between broad and loose place is particularly accentuated in Sextus: see *PH* III 119, *M* X 15 and 95. See Burnyeat [43] for Sextus on place. Annas and Barnes [8], 175 n., say that 'Sextus presumably allows that things have places in the loose sense, a sense accepted by common sense and not invented by the Dogmatists.' What Aristotle does is try to show how our practice of talking of broad places depends on the existence of proper places.

[87] 212^{a}19–20.

inner limit of an immobile surrounder; this would be nonsense if he meant to claim that the whole river was the place of the boat.

Aristotle is saying that the whole river can locate the boat, unlike the water which composes it at any one time. It is a place of the boat (not *the* place; just *a* place) just so long as it locates the boat. We can refer to it to say where the boat is. The river is a derivative place of the boat, whereas, just so long as the river locates the boat, the inner limit of the whole river where it meets the boat is the primary or proper place of the boat.

The examples of places Aristotle gives in the *Categories*—the Lyceum and the agora—are surely immobile as well. Waiving for the moment the worry one might have that they are not quite like places in the strict sense of *Physics* IV 4, we can see that, like the river, they are immobile in senses (2) and (3). It is not hard to see that being immobile in sense (2) is going to be an advantage for prospective intermediate places. Suppose you know where the Lyceum is, then if someone says that Coriscus is in the Lyceum, you are well placed for being able to find him, since the Lyceum does not move. However, if someone says that he is in the ship *Argo*, this will only be helpful if you are up to date with the *Argo*'s movements.

Indeed, the very existence of immobile objects is what enables us to say where things are. As Russell says, our conception of place comes about in large part because 'most objects on the earth's surface are fairly persistent and nearly stationary from a terrestrial point of view', and 'the whole notion that one is always in some definite "place" is due to the fortunate immobility of most of the large objects on the earth's surface'.[88] Russell concludes: 'The idea of "place" is only a rough practical approximation: there is nothing logically necessary about it, and it cannot be made precise.'[89] However, Aristotle, unlike Russell, thinks that the immobility of objects on earth, whether they be large or otherwise, is far from being 'fortunate'; they are mostly at rest *naturally*, and their immobility can be established through the immobility of the universe.

[88] Russell [139], 11, 12.

[89] Ibid. 12. It is not clear to me why Russell draws this conclusion. As a believer in relative motion, he should think that the concept 'relative to the earth' is perfectly precise, so this cannot be the problem. Rather, it seems that it is because the immobility in question is *fortunate* that the notion of place cannot be logically necessary and precise. But this does not follow.

Hence according to Aristotle, it *is* possible to be 'definite' about where things are.

Relaxing and refining the account

There is a certain amount of relaxation in Aristotle's account as far as intermediate places go, for they are not going to be immobile in quite the way that the universe and proper places are immobile. However, there is a further element of relaxation in what Aristotle says. He implies that the river water and the river surround the boat on the river. But presumably a good part of the boat is *above* the waterline and so there is no inner limit of the river which *completely* surrounds the boat, since the upper part of the boat is not surrounded by the river at all. Many things are surrounded not by a simple body, but by a complex one (including Aristotle's favoured example to illustrate the relation between something and its place: some liquid in a vessel).[90] What completely surrounds the boat is not the river, but the river and the air.

Aristotle never once in *Physics* IV 1–5 discusses examples of complete surrounding, such as a fish in water. None the less, I have argued that his concept of place relies fundamentally on the idea that *x*'s proper place is the inner limit of *x*'s surroundings, where this is understood in the first place as being the inner limit of the universe where it is in contact with *x*—then it is a question of specifying that limit as the inner limit of those things which locate *x*. However, Aristotle considers it adequate to specify this limit partially. So, for instance, the inner limit of the river where it is in contact with the boat does not account for the limit of the air where it is in contact with the boat.

There does not seem to be any danger in being loose in this way, for the following reason. As long as the river is immobile, and the boat is on the river, then it suffices that the boat be 'surrounded' by a different limit of the river for it to have changed place. Hence, we can refer ourselves to the inner limit of the river as the proper place of the boat, even though this is being slightly loose. Part of the reason why Aristotle prefers to be loose in this way is surely that if he were to specify what surrounds the boat completely, the answer would be an assemblage of air and water—however, when

[90] See IV 1–5, *passim*.

asking where the boat is, surely no one would think of relating it to the air. Boats are on rivers, and if we want to be more precise about where they are, we go on to say on which part of the river, and so on. The air does not normally enter into our practice of saying where a boat is.

It is interesting to compare this with Aristotle's remark that we are in the air in virtue of being in the surface of air which surrounds each of us;[91] strictly speaking we are surrounded by an assemblage of air and earth. Aristotle implies this at the beginning of IV 2, when he says that I am in the air and on the earth. He appeared to be saying that both these things count as my places. If we think of the limit of the earth where it is in contact with me, then we can refer to that as being my place—if I go somewhere else then one thing of which we can be sure is that I shall no longer be exactly on that same part of the earth. Notice, however, that this is still imprecise. '*x* is above such-and-such a part of the earth' does not suffice to say exactly where something is.[92]

A further relaxation of what Aristotle says is hinted at in the curious sentence 'Just as a vessel is a movable [*metaphorētos*] place, so is a place an immobile [*ametakinēton*] vessel.'[93] Charity is needed in order to avoid the obvious inconsistency;[94] I suggest that Aristotle is saying that vessels are non-strict places, because they can, and often do, move (indeed, one of the advantages of a vessel is that it can be easily moved around whereas its contents, outside of the vessel, cannot). We often say where something is by saying in what vessel it is. Particularly obvious examples of this are in biology or medicine, and indeed Aristotle himself talks of the parts of animals and plants as 'places',[95] and the Hippocratic tract *On the Places of Man* uses 'place' in this sense. These parts derive movement from their hosts, and are hence not immobile; they are not places in a strict sense—Aristotle's remarks here in the *Physics* are explicit, and places need to be immobile in order to satisfy the exigency that something which moves from being somewhere to

[91] *Phys.* IV 4, 211ᵃ26.

[92] See the Introduction, p. 7.

[93] IV 4, 212ᵃ14–16.

[94] μεταφορητός and ἀμετακίνητον need to be interpreted with the same modal force, so Aristotle is committed to either (1) places are unmoving moving vessels, or (2) places are unmovable movable vessels.

[95] See Bonitz s.v. τόπος, 767ᵃ2 ff. Typical examples of τόποι in the body are the uterus, head, intestines, bladder, chest, etc.—things *in* which gases, fluids, and solids collect.

being somewhere else must change place. Anything which is in a vessel has a proper and strict place, because it has limits. But it is a natural *relaxation* of the notion of a place to specify the container of something when saying where something is.

If specifying the vessel in which something is counts as giving its place—in a loose sense of 'place'—this might have something to do with the following. Suppose I pass you a jug of wine. To establish whether the wine is in the same place after the end of the action as it was before the start, we need to find something which locates it throughout that period.[96] Clearly, the jug will not do: it was moving too. We already know that the wine's maximal surrounder—its surroundings in the broadest sense, i.e. the universe—locates it. But is there not part of its surroundings which locates the wine as it moves? If we turn our attention to the jug, then one possible answer emerges. For if we ask what locates the jug while I pass it to you, then the answer is the air of the room, because the air of the room can be treated as being immobile although made up of air which moves, rather as a river is immobile even though its constituent water is in motion. Suppose we then consider the collection or assemblage of the jug and the air in the room. (Given that Aristotle's theory seems to demand, in its strictest applications, the notion of a collection or assemblage of items, it seems that we are justified in this.) This entity—a body—is immobile in an appropriate sense during the period in question, as is clear if we think of its outside, which is just the outer limit of the air of the room. Hence, this complex body locates the wine, and its inner limit is the place of the wine. So that which locates the wine includes that which locates the jug; perhaps this is one reason why, in a loose sense, the jug is a place of the wine.[97]

Aristotle is particularly interested in the *Physics* in the natural motions of natural bodies. This means that our refinements and relaxations will not always be necessary: for instance, the motion of a stone through the air is easily counted as just a change in limit where the air is in contact with the stone. In general, the motions which interest Aristotle are through a single fluid medium,[98] perhaps air

[96] For this principle see above, pp. 158–9.

[97] Notice how this looseness is different from the way in which *broad* places are loose places, even if there is one similarity, namely that the occupant can move and still find itself in the same place.

[98] Inwood [60], 280, goes some way towards seeing this, when he talks of 'the

or water, and these sorts of motions are well accommodated to his account. For the elements air and water are immobile—as scattered objects, they stay immobile, even if their parts are constantly in motion (notably towards the rest of the element itself). What is more, Aristotle wants to explain how it is that the nature of something's surroundings might make a difference to its behaviour. In general, *to periechon* plays a role in the behaviour of things.[99] Aristotle has shown that something's more immediate surroundings can locate it during its motion, and at the very end of his account of place[100] he mentions (and he admits that it is obscure),[101] that the surroundings of *x*, if of a different kind from *x*, will act on *x*. So, for instance, some air in water is acted on by the water in some way with the result that it rises. By linking the notion of a place to that of *to periechon*, Aristotle can explain why it is that where something is makes a difference to how it behaves.[102]

Objections

There are further objections to this line of interpretation of Aristotle. For instance, Sorabji argues that in the case of the moving boat on the river, the inner limit of the river where it touches the boat is moving, for it moves with the boat.[103] If we attempt to combat this by saying that the inner limit of the river is like a shape and merely something assumed by the river, then, he says, the surface (or limit) will be an 'instantaneous entit[y]'.[104] The boat could never return to a place, since the place only existed instantaneously.

The objection must fail. The limit of the river where it is in contact with the boat is like the shape or size of the river; if the objection holds, then it would be logically impossible for the river ever to have the same shape or size twice over. The colour of the river changes when there is an oil-slick: but this just means that the river changes in respect of its colour; surely it could perfectly well

fluids which Aristotle set up as the primary place of objects in motion' (although the reference to primary places must be wrong here).

[99] See the passages cited above, p. 135.

[100] IV 5, 212b29–213a11.

[101] 213a5–6: *ἀλλὰ διὰ τὸν καιρὸν ἀνάγκη μὲν εἰπεῖν, ἀσαφῶς δὲ νῦν ῥηθὲν τότ' ἔσται σαφέστερον* ('These questions must be determined later; while it is necessary because occasion arises to say this, what is at present obscurely expressed will then be clearer'). Translation from Hussey, [5].

[102] See Chapter 1, p. 52.

[103] Sorabji [26], 189.

[104] Ibid. 190.

regain its original colour. The objection suggests that *the surface* of the river *moves with* the boat: but in fact, the *river* just *changes in respect of its form* around the boat as it moves; surely it could perfectly well regain the same form around that boat or another boat. Shapes, sizes, limits, colours, etc. are assumed—sometimes instantaneously—and they can be reassumed. Our interpretation is safe from this objection.

The major difficulty that interpreters have had with Aristotle is attaching a sense to how the internal limit of the river around, say, a stationary boat could possibly be considered *immobile*. Burnyeat's suggestion that we conceive the river as a geographical entity contained the key to a proper understanding of this. Even though the inner limit of the river around the boat changes materially or in respect of what composes it, it is the same limit according to our conditions of individuation, for the river as a whole keeps the same shape and size around the boat, as long as the river itself does not move. Aristotle's account commits him to this type of account of the individuation of limits, since he himself is prepared to accept the inner limit of the river *qua* whole river as remaining the same, even if materially or compositionally speaking this limit is changing constantly.

There is, however, a powerful objection to Aristotle's account, advanced by Owen.[105] Aristotle says that those things which locate x must be immobile. Then their inner limit can be considered as the place of x. But 'immobile' simply means 'not changing place', in which case Aristotle's definition of place is circular. Aristotle cannot, it seems, elucidate what a place is without making reference to something which does not change place.[106] The objection is a good one, for it forces us to acknowledge one important feature of Aristotle's account. It is not necessary to explain the immobility of the maximal surrounder of something by adverting to its *place*. It is immobile because it is impossible for it to move: it can displace nothing. There is nothing outside of it relative to which it can move. (In consequence, we might say, there is nowhere else for it to go.) It is not 'immobile' in the sense of having somewhere to go but not going, or being able to go, there. Thus, bodies within

105 [76], 252.

106 This consideration is one of those which lead Hussey, [5], 117, to conclude that the final definition of place which includes the predicate 'immobile' is an interpolation.

the universe are immobile in virtue of the fact that their maximal surrounder retains the same shape around them. *This* does not introduce circularity. Rather, it anchors the notion of absolute local movement and location to the finite Aristotelian universe and its plenitude. The immobility of those things which locate a given object, i.e. those things which serve as a place for that object, can also be explained in this way.

Rotation and the place of the universe

One difficulty that has often surfaced in discussions of Aristotle's account of place is the problem of the place of the universe. On the one hand, Aristotle says explicitly that the universe has no place, since there is nothing outside of it. On the other, he clearly thinks that it is somewhere (after all, everything is somewhere). I have already alluded to how the universe is somewhere in virtue of being in its parts, which all have places.[107] None the less, there are a whole host of problems here which merit attention, and which together can constitute a sort of 'test case' for Aristotle's account of place.

Let us start with the claim that the universe is somewhere in virtue of being in its parts. This only works if all the universe's parts are themselves P-items—items which are in a place. However, talking of the universe's parts is vague. There are many ways to partition up the universe into a set of parts, and it is clear that not all such partitions will be sets of P-items. For instance, suppose I partition up the universe into the sphere of the earth, the sphere of water, the sphere of air, the sphere of fire, and the celestial sphere. This will not be a partition such that all its members are P-items, for the celestial sphere—i.e. the outer sphere of the universe—is not a P-item. (There is nothing outside *it*.)

There are, however, many partitions of the universe which are sets of P-items (in fact, there are infinitely many), e.g. the set consisting of the four sublunary spheres plus the outer sphere segmented up into any number of segments. Each segment of the outer sphere, although on the edge of the universe, has a specifiable place within the universe—it would make a recognizable 'hole' in the universe. (It would not quite be right to talk of the *places* of these parts, but rather their *potential* places, i.e. the places they would have were

[107] See Chapter 3, p. 99.

they to be separate from their surroundings rather than continuous.[108]) The segments will not be *totally* surrounded by something, but they will be for the most part.[109] So the universe is somewhere in respect of its parts in virtue of the fact that (1) it is not in a place, and (2) there is a partition of it such that all members of that partition are P-items. Notice also that the outer sphere of the universe, although not a P-item, has also been shown in the process to be somewhere (it is in its parts, and there is a partition of it such that its parts are all P-items).

So much for the lack of place of the universe, and the outer sphere of the universe. However, Aristotle clearly maintains that the heavens rotate, and he seems to hold that rotation is a case of local motion.[110] But the worry is: how can something indulge in local motion if it has no place?

To answer this, it is crucial to start with consideration of a rotating object which *does* have a place, e.g. a spinning top. As was noted by Plato,[111] such a top could be spinning but not changing place. Not all cases of rotation are like that: for instance, a rotating cube *is* changing place, and in this instance rotation is straightforwardly a change of place. The top, however, rotates and does not change place. Its parts are in motion—are changing place—but not the whole. *It* moves in respect of its parts. Again, we have to ask what we mean when we say 'its parts'. For it is not the case that on every partition of the top, its parts will be changing place. Imagine, for instance, that we cut the top in two horizontally into an upper half and a lower half. Neither of those two halves is changing potential place (both are spinning in the same place). But there are partitions of the top such that, on that partition, its parts are changing potential place, for instance if we partition the top into two halves vertically, to give a front half and a back half.

[108] This means that we shall have to stretch the notion of a P-item to include items with *potential* places. Notice that this does not fall foul of the objection that upset the theory of space considered in Chapter 4, pp. 121–32. There, the problem was that the parts of a continuous whole (e.g. some water) have a place *in exactly the same way as* the whole has a place; but in the case under consideration here, the parts of a continuous whole have *potential* places (not *actual* places), mirroring the fact that the parts of a continuous whole exist in some *potential* sense.

[109] I have already shown how Aristotle's account of place can be extended to cope with cases of partial surrounding—indeed, his standard examples of water in a jug, and the boat on the river, are such cases.

[110] All the evidence for this is admirably collected by Simplicius at the start of his Corollary on Place (*In Phys.* 601. 1–645. 19).

[111] *Rep.* 436 D 4–E 6.

Hence if the top is rotating then there is a partition of the top such that all the members of that partition are changing (potential) place.

We can develop this into a general definition of rotation in the following way: x is rotating iff there is a partition P of x such that all the members of P are changing potential place circularly. The condition 'circularly' is necessary because obviously something could be in a state of chaotic flux—its parts could all be in furious motion—and yet not be *rotating*, e.g. a cloud. I shall not attempt to define more precisely what 'circularly' means here. Note that this definition allows for something to be rotating and also moving rectilinearly (as in fact most tops do when they spin). It can also be applied to cases where an object is rotating and this does in fact entail a change of place, e.g. when a cube rotates.

Let us now turn back to the case of the universe. If the outer sphere of the universe is rotating, then there must be a partition of it such that the members of the partition are changing place circularly. Well, suppose we segment up the outer sphere as before. The segments are P-items. If the outer sphere is rotating, then these parts will all change place circularly. There is no difficulty that the outer sphere has no place, because rotation is obviously change of place in respect of the parts of something—what is important is that the outer sphere can be partitioned up into items which do have places, or potential places. *But this is something which we have to do anyway in order to explain how it is that the universe is somewhere.* Therefore Aristotle is perfectly justified in holding the following propositions to be true: (1) the universe has no place; (2) the outer sphere of the universe has no place; (3) the universe is somewhere; (4) the outer sphere of the universe is somewhere; (5) the outer sphere of the universe is rotating. A proper account of rotation recognizes that all that is necessary for something to rotate is that its parts be P-items—be the sorts of thing which have a place or a potential place, which they can change circularly. The outer sphere of the universe is such an item.

The point of concentrating on the case of the universe and rotation is to show how Aristotle's concept of place can be applied to these theoretical issues and help explain them. It is true that a certain amount of work is needed to render perfectly acceptable the account of something's being somewhere in respect of its parts. But I have already given an example of how philosophers need such a notion anyway.[112] Aristotle's account of place can, I believe, meet the objection that according to him the outer sphere of the universe has no place, and yet rotates.

Relative and absolute place

The *Physics* is supposed at least in part to be an exercise in the clarification of certain key concepts which are used in the natural sciences. Place is among these key concepts, because of its intimate relationship with motion, which is in turn exactly what differentiates natural bodies from non-natural ones; natural ones have within them a principle of motion and rest. Motion is of importance in astronomy, in which the laws of motion of the stars and planets are to be discovered, and in biology, since it is typical of animate bodies that they undergo changes due to their having a soul, and among these changes are local motions such as walking, running, swimming, crawling, flying, etc. These last movements in particular are conspicuously movements *through* certain media, and it is clear also from Aristotle's remarks on replacement (*antimetastasis*) and *antiperistasis* that his model for motion is motion through a medium entailing displacement. Further, I have argued that Aristotle's talk of *to periechon* throughout his discussion of place is to be taken, in the first instance, as a reference to the entire universe, which is to be understood as a sort of medium through which mobile bodies move. Given this approach to his subject, it is not surprising that Aristotle does not do battle with the question of relative and absolute place. Movement is, for Aristotle, movement *through* some *external* body or group of bodies. Whether or not a given body is engaged in motion may then seem to depend on which external body or group of bodies you take as your reference-point. However, since the universe has no body external to it through which *it* can move—nowhere to go, as it were—and since it is external to every body it contains, it is the obvious body (or group of bodies) to be the reference-point. Once this is granted, the theory of intermediate places and proper places can be derived.

For these reasons, Aristotle thinks that there is an answer to the question 'is *x* moving or not?' without it being necessary to ask further 'relative to what?' After all, if someone asks whether

[112] See Chapter 3, p. 99 n. 58.

such-and-such a bird is in flight, or such-and-such a fish is swimming, or such-and-such a human is running, we do not need to ask relative to what it is flying, swimming, or running. In this sense, Aristotle is obviously a partisan of absolute motion and place. The impossibility of the universe moving anywhere guarantees this absolutism. As Eudemus said: 'something which moves is like a vessel, so we refer places to the heavens. For these do not ever change place'.[113] However, Aristotle's theory of absolute place is very different from classical—Newtonian—absolute place, and indeed has much in common with the classical relative picture of place. It is worth touching on this, although the subject is clearly too large for full treatment here.

Advocates of absolute space tend to identify space with a certain substance, parts of which are occupied by bodies and which are their (absolute) places. This corresponds to what Reichenbach calls 'the crude conception of a substantial space which Newton shares with the epistemologically untutored layman'.[114] But of course, Aristotle has no such conception of space (indeed, he has no conception of space),[115] and certainly that is not his conception of place. The closest Aristotle gets to such a conception is the interval theory of place, which he strongly rejects.[116] The Aristotelian place of a body, x, is to be identified with bodies other than x, or features (surfaces) thereof. Thus Aristotle's absolutism is not to be compared with classical absolute theories of space and place.

Conceptions of place as relative also deny, as Aristotle does, that places are substances over and above bodies. According to theories of relative place, something is in the same place as something else when the totality of spatial relations they bear to other bodies or a certain group of bodies are the same. (Leibniz, for instance, held this type of theory.) But on this theory, places are not even items in the world, unlike Aristotle's conception, so it would be highly misleading to call Aristotle's conception of place relative. Presumably here again, Aristotle has common sense on his side, for we

[113] Fr. 80 Wehrli: *τὸ γὰρ κινούμενον ἀγγειῶδες, καὶ διὰ τοῦτο τῶν τόπων τὴν ἀναφορὰν πρὸς τὸν οὐρανὸν ποιούμεθα. οὗτος γὰρ οὐ μεταλλάσσει τόπον ἄλλον.* Here I follow up Sedley's remark ([65], 140 n. 1) that Eudemus must have had the same conception of place as Aristotle.

[114] Reichenbach [138], 58.

[115] Bergson was the first to see this. See Bergson [42], 74: 'Habemus igitur principium metaphysicum, ex quo progressus Aristoteles ad infitiationem nostri quod dicimus spatii pervenerit'; and *passim*.

[116] See Chapter 4, pp. 121–32.

certainly refer ordinarily to places which are concrete items in the world (land masses, islands, houses, valleys, rock pools, etc.).

However, Aristotle clearly thinks that in order to say *where* a body is, that body needs to be related to *other* bodies, and this is similar to the relative conception, doubtless explaining why Weyl [130], 151, says: '[It] is not true . . . that the conception of rest has an objective content', and then adds in a note: 'Even Aristotle was clear on this point, for he denotes "place" (τόπος) as the relation of a body to the bodies in its neighbourhood.' However, I have argued that although Aristotle's definition of place involves *relating* a body to its surroundings, not any statement to the effect that such-and-such a body is related to such-and-such another body will do. For Aristotle, these other bodies have to fulfil certain conditions, namely, they have to surround the original body and be immobile; furthermore, they can then be *identified* with the body's (intermediate) places. The surroundings (*to periechon*) to which a body must be related in order to say where it is are the whole universe; but these surroundings are immobile—at absolute rest—and hence, in *relating* a body to its surroundings in the right way, it *is* possible to say *absolutely* where it is. Whether this means that Aristotle's concept of place is relative or absolute seems to me an empty question.

Conclusion

My interpretation of Aristotle's theory of place has affinities with the medieval concept of formal place, which was introduced in order to combat perceived difficulties in Aristotle's account, notably the problem that if something's place is the inner limit of its surroundings, then that inner limit could be changing (e.g. if it is the inner limit of something moving like air or water) while the surrounded object was stationary.[117] However, my interpretation diverges from that theory in one important respect: the proper place of something, according to me, is an actual item—a surface—whereas according to the theory of formal place it is not. Simply, it is clear that Aristotle had the idea that what surrounds something can be specified in different ways, ways which make a difference to whether the inner limit of what surrounds is immobile or not.

[117] See Grant [68], 63–72, with Sorabji [26], 190, and Algra [56], 145 n. 10 and [21], 252–4.

(This is why Aristotle distinguishes between picking out the inner surface of the boat's surroundings as the limit of the river *water*, or as the limit of the river.) It makes a difference how you specify the inner limit of something's surroundings; only some ways will count as specifications of the place of that thing. This is why we need the notion of intermediate, or broad, places. And, importantly, although the sense in which the inner limit of the river remains the same around the boat is to be elucidated in terms of shape and size—i.e. formal features of the river—the place of the boat is not *identified* with that form. My interpretation of Aristotle takes what he says at face value and does not introduce a distinction between material and formal place.

Throughout, I have had an eye on where-questions. Although Aristotle never once mentions such questions in the *Physics*, his account starts with the observation that everyone holds that everything that is is somewhere. But for all things *x*, if *x* is somewhere, then there is somewhere where *x* is—and if there is somewhere where *x* is, then for any *x*, the question arises: where is *x*? There is thus an obvious link between something's being somewhere and our practice of asking where it is.[118] However, at first glance it might seem as if Aristotle's account of place cannot possibly be a contribution to an understanding of *how* we *answer* where-questions. When do we *ever* answer 'where is *x*?' by specifying a particular immobile surface?

This is to approach Aristotle's account from the wrong angle. For a start, what counts as an adequate answer to a question is a difficult matter. Adequate answers depend on the interests of the questioner, the capacities of the respondent, and a whole host of other contingent factors. Moreover, it is clear that we ask where-questions about things that we would not normally say have places (e.g. properties). Even for those items which do have places, 'where is *x*?' might be answered perfectly adequately and quotidianly by '[*x* is] to the right of *y*', but such answers cannot tell the *whole* story about our practice of locating things, because the questions 'where is *a*?' and 'where is *b*?', where *a* and *b* are *not* in the same place, might provoke the *same* answer 'to the right of *c*'.[119] Hence, it is not obvious that considering all

[118] This is obvious in Greek, since the words for 'somewhere' and 'where?' differ only in accentuation.

[119] See the Introduction, p. 7.

the different answers we give to where-questions will reveal what places are.

Aristotle's aim seems rather to have been to articulate the underlying assumptions in our practice of saying where things are by saying what they are *in*; he shows that this practice implies the existence of *proper* places, which are in turn needed to elucidate principles such as 'motion is change of place', 'two bodies cannot be in the same place at the same time', and 'one body cannot be in two distinct places at the same time'. Because saying what something's surroundings are (saying what it is *in*) is the canonical way to answer someone who asks *where* it is, Aristotle can conclude that this is what it is for things with surroundings to be *somewhere*. But many different types of thing are somewhere, and so this notion needs analysis; the analysis he succeeds in giving is a philosophical achievement which ranks alongside the analysis in the *Metaphysics* of the notion of being. Aristotle takes the bull by the horns and defines a strict notion of being somewhere along with the items to which it applies, namely, those bodies which are surrounded by other bodies (P-items). P-items serve as the basis for saying where all other items are, such as places themselves, other properties, the parts of things, and the universe, as well as other items which Aristotle does not mention, like causes, sounds, and shadows.

What he says connects in this way with where-questions: we say of many things that they are somewhere, and hence we ask of many items where they are. 'Where is *x*?', where *x* is a P-item, can be answered by specifying its broad place, or a non-strict (mobile) place, if that is what the questioner wants, *but things have broad or non-strict places only if they are P-items*. Similarly, where-questions concerning things which are not P-items will typically be answered by referring *to* P-items. But a P-item is somewhere in virtue of being in its proper place. Hence, even if strict proper places are not what *feature in* answers to where-questions, they are none the less the key to an understanding of what it is to ask, and how it is possible to answer, where-questions—how it is possible to say *where* things are, or come to know *where* they are. Hence Aristotle provides no less than the metaphysical underpinning necessary for an understanding of what we mean when we say that something—or, indeed, everything—is *somewhere*.

Bibliography

This bibliography comprises all texts, books, and articles to which I have referred explicitly, and some other works which could not be left unmentioned. The only modern book-length studies dedicated exclusively to Aristotle's concept of place are Bergson [42] and Zekl [55]. The best starting-point in the secondary literature is Burnyeat [43]. I have included some references to modern treatments of the notion of place.

I. TEXTS

I have used standard editions throughout.

II. TRANSLATIONS AND COMMENTARIES

Aristotle

[1] BARNES, J. (ed.), *The Complete Works of Aristotle* (Princeton, 1982).

[2] WARDY, R., *The Chain of Change: A Study of Aristotle's* Physics *VII*, text, translation, and commentary (Cambridge, 1990).

[3] ROSS, W. D., *Aristotle's* Physics, text with introduction, commentary, and analysis (Oxford, 1936).

[4] WATERFIELD, R. (trans.), *Aristotle:* Physics, with introduction and notes by D. Bostock (Oxford, 1996).

[5] HUSSEY, E., *Aristotle's* Physics *Books III and IV*, translation with notes (Clarendon Aristotle Series; Oxford, 1983; rev. edn. 1993).

[6] CHARLTON, W., *Aristotle's* Physics *Books I and II*, translation with notes (Clarendon Aristotle Series; Oxford, 1970; rev. edn. 1992).

[7] GRAHAM, D. W., *Aristotle* Physics *Book VIII*, translation with notes (Clarendon Aristotle Series; Oxford, 1999).

Other Authors

[8] ANNAS, J., and BARNES, J., *Sextus Empiricus:* Outlines of Scepticism (Cambridge, 1994).

[9] BARNES, J., *Early Greek Philosophy* (London, 1987).

[10] CORNFORD, F. M., *Plato and Parmenides* (London, 1939).

[11] —— *Plato's Cosmology* (London, 1937).

[12] FURLEY, D., 'Philoponus: Corollaries on Place and Void', in D. Furley

and C. Wildberg, *Philoponus: Corollaries on Place and Void, with Simplicius: Against Philoponus on the Eternity of the World* (Ancient Commentators on Aristotle; London, 1991), 7–94.

[13] Gottschalk, H. B., *Strato of Lampsacus: Some Texts* (Proceedings of the Leeds Philosophical and Literary Society, Literary and Historical Section, 11/6: 95–182; Leeds, 1965).

[14] Heath, T., *The Thirteen Books of* The Elements, 2nd edn. (3 vols.; Cambridge, 1956).

[15] Sharples, R., *Theophrastus of Eresus: Commentary*, iii/1. *Sources on Physics* (Philosophia Antiqua, 79; Leiden, 1998).

[16] Strange, S. K., *Porphyry: On Aristotle* Categories (Ancient Commentators on Aristotle; London, 1992).

[17] Taylor, A. E., *A Commentary on Plato's* Timaeus (Oxford, 1928).

[18] Urmson, J. O., *Simplicius: On Aristotle* Physics *4. 1–5, 10–14* (Ancient Commentators on Aristotle; London, 1992).

[19] —— *Simplicius: Corollaries on Place and Time* (Ancient Commentators on Aristotle; London, 1992).

[20] —— 'Simplicius: On Aristotle on the Void', in P. Lettinck and J. O. Urmson, *Philoponus: On Aristotle* Physics *5–8, with Simplicius: On Aristotle on the Void* (Ancient Commentators on Aristotle; London, 1994), 167–236.

III. PLACE, SPACE, AND COSMOLOGY IN GREEK THOUGHT

General

[21] Algra, K., *Concepts of Space in Greek Thought* (Philosophia Antiqua, 65; Leiden, 1994).

[22] Duhem, P., *Le Système du monde*, i (Paris, 1913).

[23] Heath, T., *Aristarchus of Samos: The Ancient Copernicus* (Oxford, 1913).

[24] Jammer, M., *Concepts of Space* (Cambridge, Mass., 1954).

[25] Sambursky, S., *The Physical World of the Greeks* (London, 1956).

[26] Sorabji, R., *Matter, Space and Motion* (London, 1988).

Presocratics

[27] Barnes, J., *The Presocratic Philosophers*, 2nd edn. (London, 1982).

[28] Cornford, F. M., 'The Invention of Space', in J. A. K. Thomson (ed.), *Essays in Honour of Gilbert Murray* (London, 1936), 215–35; repr. in *Selected Papers of F. M. Cornford*, ed. A. C. Bowen (Greek and Roman Philosophy, 10; London, 1987), 199–219.

[29] Kahn, C. H., *Anaximander and the Origins of Greek Cosmology* (New York, 1960).

[30] KEMBER, O., 'Right and Left in the Sexual Theories of Parmenides', *Journal of Hellenic Studies*, 91 (1971), 70–9.

[31] LLOYD, G. E. R., 'Right and Left in Greek Philosophy', *Journal of Hellenic Studies*, 82 (1962), 168–79; repr. in his *Methods and Problems in Greek Science* (Cambridge, 1991), 27–48.

[32] —— 'Parmenides' Sexual Theories: A Reply to Mr Kember', *Journal of Hellenic Studies*, 92 (1972), 178–9.

Plato

See CORNFORD [11], TAYLOR [17].

Aristotle

(**a**) Natural place

[33] BARR, R., 'Aristotle on Natural Place: Some Questions', *The New Scholasticism*, 30 (1956), 206–10.

[34] KRONZ, F. M., 'Aristotle, the Direction Problem, and the Structure of the Sublunar Realm', *The Modern Schoolman*, 67 (1990), 247–57.

[35] MACHAMER, P. K., 'Aristotle on Natural Place and Natural Motion', *Isis*, 69 (1978), 377–87.

[36] WEISHEIPL, J. A., 'Space and Gravitation', *The New Scholasticism*, 29 (1955), 175–223.

[37] —— 'Aristotle on Natural Place: A Rejoinder', *The New Scholasticism*, 30 (1956), 211–15.

(**b**) Aristotle on Plato on place and space

[38] CHERNISS, H., *Aristotle's Criticism of Plato and the Academy*, i (Baltimore, 1944).

[39] CLAGHORN, G. S., *Aristotle's Criticism of Plato's* Timaeus (The Hague, 1954).

[40] KEYT, D., 'Aristotle on Plato's Receptacle', *American Journal of Philology*, 82 (1961), 291–300.

See also ALGRA [21], § 3.3; and appendix 2, 'Aristotle's Doctrine of Space', to TAYLOR [17], 664–77.

(**c**) Place in the *Categories*

See ACKRILL [83], 91–98; ALGRA [21], 123–36; MENDELL [50].

(**d**) Place in the *Physics*

[41] BARREAU, H., 'L'espace et le temps chez Aristote', *Revue de métaphysique et de morale*, 80 (1975), 417–38.

[42] BERGSON, H., *Quid de loco senserit Aristoteles* (diss. Paris, 1889); French translation (R. Mossé-Bastide): 'L'Idée de lieu chez Aristote', *Études bergsoniennes*, 2 (1949), 27–109. English translation

with an introduction (John K. Ryan): 'Aristotle's Concept of Place', *Studies in Philosophy and the History of Philosophy*, 5 (1970), 13–72.

[43] BURNYEAT, M., 'The Sceptic in his Place and Time', in R. Rorty, J. B. Schneewind, and Q. Skinner (eds.), *Philosophy in History* (Cambridge, 1984), 225–54; repr. in M. Burnyeat and M. Frede (eds.), *The Original Sceptics: A Controversy* (Indianapolis, 1997), 92–126.

[44] CARTWRIGHT, H., 'Parts and Places', in J. J. Thomson (ed.), *On Being and Saying: Essays in Honor of Richard Cartwright* (Cambridge, 1987), 175–214.

[45] DEHN, M., 'Raum, Zeit, Zahl bei Aristoteles vom mathematischen Standpunkt aus', *Scientia* (Bologna), 60 (1936), 12–21, 69–74; repr. in [77], 199–218.

[46] GOLDSCHMIDT, V., 'La théorie aristotélicienne du lieu', in *Mélanges de philosophie grecque offerts à Mgr. Diès* (Paris, 1956), 79–119.

[47] KING, H. R., 'Aristotle's Theory of Topos', *Classical Quarterly*, 44 (1950), 76–96.

[48] LANG, H., *The Order of Nature in Aristotle's Physics: Place and the Elements* (Cambridge, 1998).

[49] MARIÑA, J., 'The Rôle of Limits in Aristotle's Concept of Place', *Southern Journal of Philosophy*, 31 (1993), 205–16.

[50] MENDELL, H., 'Topoi on Topos: The Development of Aristotle's Concept of Place', *Phronesis*, 32 (1987), 206–31.

[51] MOREAU, J., 'L'espace chez Aristote', *Giornale di metafisica*, 4 (1949), 351–69, 525–42.

[52] —— *L'Espace et le temps selon Aristote* (Padua, 1965).

[53] MORISON, B., review of [48], *Classical Review*, 50/2 (2000), 493–5.

[54] SESMAT, A., 'La théorie aristotélicienne du lieu', *Revue de philosophie*, 38 (1938), 477–500.

[55] ZEKL, H. G., *Topos: Die aristotelische Lehre vom Raum: Eine Interpretation von* Physik, *D 1–5* (Paradeigmata, 10; Hamburg, 1990).

See also ALGRA [21], ch. 4; Bostock in WATERFIELD [4], xxxvi–xl; HUSSEY [5], xxvii–xxxii; OWEN [75], 92–5, and [76], 252–3; ROSS [3], 53–8; and SORABJI [26], 186–92.

Peripatetic places

[56] ALGRA, K., '"Place" in Context: On Theophrastus Fr. 21 and 22 Wimmer', in W. W. Fortenbaugh and D. Gutas (eds.), *Theophrastus: His Psychological, Doxographical and Scientific Writings* (Rutgers University Studies in Classical Humanities, 5; London, 1992), 141–65.

[57] FURLEY, D., 'Strato's Theory of the Void', in J. Wiesner (ed.), *Aristoteles: Werk und Werkung*, i. *Aristoteles und seine Schule* (Berlin, 1985), 594–609.

[58] RACHED, M., 'Alexandre d'Aphrodise et la "magna quaestio"', *Études classiques*, 63 (1995), 295–351.

[59] SORABJI, R., 'Theophrastus on Place', in W. W. Fortenbaugh and R. W. Sharples (eds.), *Theophrastean Studies: On Natural Science, Physics and Metaphysics, Ethics, Religion, and Rhetoric* (Rutgers University Studies in Classical Humanities, 3; London, 1988), 139–66.

See GOTTSCHALK [13], appendix 2, for Strato of Lampsacus; *see also* ALGRA [21], ch. 5; and SORABJI [26], ch. 11.

Epicurean places

[60] INWOOD, B., 'The Origin of Epicurus' Concept of Void', *Classical Philology*, 76 (1981), 273–85.

[61] SEDLEY, D., 'Two Conceptions of Vacuum', *Phronesis*, 27 (1982), 175–93.

Sceptic places

See BURNYEAT [43].

Stoic places

See ALGRA [21], ch. 6, with bibliographical references.

Later places

[62] HOFFMANN, P., 'Simplicius: Corollarium de loco', in *L'Astronomie dans l'antiquité classique: Actes du colloque tenu à l'Université de Toulouse-le-Mirail, 1977* (Paris, 1979), 143–63.

[63] SAMBURSKY, S., *The Physical World of Late Antiquity* (London, 1962).

[64] —— *The Concept of Place in Late Platonism* (Jerusalem, 1982).

[65] SEDLEY, D., 'Philoponus' Conception of Space', in R. Sorabji (ed.), *Philoponus and the Rejection of Aristotelian Science* (London, 1987), 140–53.

[66] VERBEKE, G., 'Ort und Raum nach Aristoteles und Simplikios: Eine philosophische Topologie', in J. Irmscher and R. Müller (eds.), *Aristoteles als Wissenschaftstheoretiker* (Berlin, 1983), 113–22.

[67] WIELAND, W., 'Zur Raumtheorie des Johannes Philoponus', in E. Fries (ed.), *Festschrift J. Klein* (Göttingen, 1967), 114–35.

See also SORABJI [26].

Medieval places

[68] GRANT, E., 'The Medieval Doctrine of Place: Some Fundamental Problems and Solutions', in A. Maierù and A. Paravicini Bagliani

(eds.), *Studi sul XIV secolo in memoria di Annaliese Maier* (Rome, 1981), 57–79.

[69] Lang, H. S., 'Aristotle's Immaterial Mover and the Problem of Location in *Physics* VIII', *Review of Metaphysics*, 35 (1981), 321–35; reappears as ch. 4 of [70].

[70] —— *Aristotle's* Physics *and Its Medieval Varieties* (New York, 1992).

IV. THE *PHYSICS*

General

[71] De Gandt, F., and Souffrin, P. (eds.), *La Physique d'Aristote et les conditions d'une science de la nature* (Paris, 1991).

[72] Judson, L. (ed.), *Aristotle's* Physics (Oxford, 1991).

[73] Le Blond, J. M., *Logique et méthode chez Aristote: Étude sur la recherche des principes dans la physique aristotélicienne* (Paris, 1939).

[74] Mansion, A., *Introduction à la physique aristotélicienne* (Louvain and Paris, 1946).

[75] Owen, G. E. L., '*Tithenai ta phainomena*', in S. Mansion (ed.), *Aristote et les problèmes de méthode* (Louvain, 1961), 83–103; repr. in [125], 239–51; also in J. M. E. Moravcsik (ed.), *Aristotle: A Collection of Critical Essays* (London, 1968), 167–90, and [91], 113–25.

[76] —— 'Aristotle: Method, Physics and Cosmology', in C. C. Gillespie (ed.), *Dictionary of Scientific Biography*, i (New York, 1970), 250–8; repr. in [125], 151–64.

[77] Seeck, G. A., (ed.), *Die Naturphilosophie des Aristoteles* (Wege der Forschung, 225; Darmstadt, 1975).

[78] Solmsen, F., *Aristotle's System of the Physical World: A Comparison with his Predecessors* (Ithaca, NY, 1960).

[79] Wieland, W., *Die aristotelische Physik*, 2nd edn. (Göttingen, 1970).

Text

[80] Barnes, J., 'Roman Aristotle', in J. Barnes and M. Griffin (eds.), *Philosophia Togata*, ii. *Plato and Aristotle at Rome* (Oxford, 1997), 1–69.

[81] Brunschwig, J., 'Qu'est-ce la *Physique* d'Aristote?', in [71], 11–40.

[82] Solmsen, F., 'Misplaced Passages at the End of Aristotle's *Physics*', *American Journal of Philology*, 82 (1961), 270–82.

See also Ross [3], 102–18.

V. OTHER SECONDARY LITERATURE REFERRED TO

[83] ACKRILL, J. L., *Aristotle's* Categories *and* De Interpretatione, translation with notes (Clarendon Aristotle Series; Oxford, 1963).

[84] AUSTIN, J. L., *Sense and Sensibilia*, ed. G. Warnock (Oxford, 1962).

[85] BARNES, J., *Aristotle:* Posterior Analytics, translation with notes (Clarendon Aristotle Series; Oxford, 1975; rev. edn. 1993).

[86] —— *Aristotle* (Oxford, 1982).

[87] —— 'Aristotelian Arithmetic', *Revue de philosophie ancienne*, 1 (1985), 97–133.

[88] —— 'Bits and Pieces', in J. Barnes and M. Mignucci (eds.), *Matter and Metaphysics* (Naples, 1988), 223–94.

[89] —— *The Toils of Scepticism* (Cambridge, 1990).

[90] —— 'Ammonius and Adverbs', *Oxford Studies in Ancient Philosophy*, suppl. vol. (1991), 145–63.

[91] —— SCHOFIELD, M., and SORABJI, R. (eds.), *Articles on Aristotle*, i. *Science* (London, 1975).

[92] —— —— —— *Articles on Aristotle*, iii. Metaphysics (London, 1978).

[93] BOSTOCK, D., *Aristotle:* Metaphysics *Books Z and H* (Clarendon Aristotle Series; Oxford, 1994).

[94] BURNYEAT, M., 'Examples in Epistemology: Socrates, Theaetetus and G. E. Moore', *Philosophy*, 52 (1977), 381–98.

[95] FREDE, M., 'The Original Notion of Cause', in M. Schofield, M. Burnyeat, and J. Barnes (eds.), *Doubt and Dogmatism* (Oxford, 1980), 217–49; repr. in [98], 125–50 (page references are to the reprint).

[96] —— 'Categories in Aristotle', in D. J. O'Meara, *Studies in Aristotle* (Studies in Philosophy and the History of Philosophy 9; Washington DC, 1981), 1–24; repr. in [98], 29–48 (page references are to the reprint).

[97] —— 'Substance in Aristotle's *Metaphysics*', in A. Gotthelf (ed.), *Aristotle on Nature and Living Things* (Pittsburgh, 1985), 17–26; repr. in [98], 72–80 (page references are to the reprint).

[98] —— *Essays in Ancient Philosophy* (Oxford, 1987).

[99] —— 'The Definition of Sensible Substances in Met. *Z*', in D. Devereux and P. Pellegrin (eds.), *Biologie, logique et métaphysique chez Aristote* (Paris, 1990), 113–29.

[100] —— 'Aristotle's Notion of Potentiality in *Metaphysics Θ*', in T. Scaltsas, D. Charles, and M. L. Gill (eds.), *Unity, Identity and Explanation in Aristotle's Metaphysics* (Oxford, 1994), 173–93.

[101] —— and PATZIG, G., *Aristoteles:* Metaphysik *Z*, text, translation, and commentary (Munich, 1988).

[102] FREGE, G., *The Foundations of Arithmetic: A Logico-Mathematical Enquiry into the Concept of Number* (Oxford, 1950); translation (J. L. Austin) of *Die Grundlagen der Arithmetik: Eine logisch mathematische Untersuchung über den Begriff der Zahl* (Breslau, 1884).

[103] —— 'On the Law of Inertia', in [106], 123–36; translation (H. Kaal) of 'Über das Trägheitsgesetz', *Zeitschrift für Philosophie une philosophische Kritik*, 98 (1891), 145–61; repr. in [105], 113–24.

[104] —— 'What is a Function?', in [106], 285–92; translation (P. Geach) of 'Was ist eine Funktion?', in S. Meyer (ed.), *Festschrift Ludwig Boltzmann gewidmet zum sechzigsten Geburtstage 20. Februar 1904* (Leipzig, 1904), 656–66; repr. in G. Frege, *Funktion, Begriff, Bedeutung*, ed. G. Patzig (Göttingen, 1962), 81–90, and in [105], 273–80.

[105] —— *Kleine Schriften*, ed. I. Angelelli (Darmstadt and Hildesheim, 1967).

[106] —— *Collected Papers on Mathematics, Logic, and Philosophy*, ed. B. McGuinness, trans. M. Black *et al.* (Oxford, 1984).

[107] FURLEY, D., 'Aristotle and the Atomists on Motion in a Void', in P. K. Machamer and R. G. Turnbull (eds.), *Motion and Time, Space and Matter: Interrelations in the History of Philosophy and Science* (Columbus, Oh., 1976), 83–100; repr. in his *Cosmic Problems: Essays on Greek and Roman Philosophy of Nature* (Cambridge, 1989), 77–90 (page references are to the reprint).

[108] GEACH, P., 'Some Problems about Time', *Proceedings of the British Academy*, 51 (1965), 321–36; repr. in his *Logic Matters* (Cambridge, 1972), 302–18 (page references are to this reprint), and in P. Strawson (ed.), *Studies in the Philosophy of Thought and Action* (Oxford, 1965), 175–91.

[109] GILL, M. L., *Aristotle on Substance* (Princeton, 1989).

[110] HIGGINBOTHAM, J., 'Interrogatives', in K. Hale and S. J. Keyser (eds.), *The View from Building 20: Essays in Honor of Sylvain Bromberger's Fiftieth Birthday* (Cambridge, Mass., 1993), 195–228.

[111] HINTIKKA, J., 'Questions about Questions', in M. K. Munitz and P. Unger (eds.), *Semantics and Philosophy* (New York, 1974), 103–58.

[112] —— 'Answers to Questions', in H. Hiz (ed.), *Questions* (Dordrecht, 1978), 279–300.

[113] HOFFMANN, P., 'Les Catégories POU et POTE chez Aristote et Simplicius', in P. Aubenque (ed.), *Concepts et catégories dans la pensée antique* (Paris, 1980), 217–45.

[114] JOACHIM, H. H., *Aristotle:* On Coming-to-be and Passing-away, revised text with introduction and commentary (Oxford, 1922).

[115] KAHN, C. H., *The Verb 'Be' in Ancient Greek*, pt. vi of J. W. M.

Verhaar (ed.), *The Verb 'Be' and its Synonyms* (Foundations of Language Supplementary Series, 16; Dordrecht, 1973).

[116] —— 'Questions and Categories: Aristotle's Doctrine of Categories in the Light of Modern Research', in H. Hiz (ed.), *Questions* (Dordrecht, 1978), 227–78.

[117] Kirwan, C., *Aristotle:* Metaphysics *Books Γ, Δ, and E*, translation with notes (Clarendon Aristotle Series; Oxford, 1971; rev. edn. 1993).

[118] Lewis, D., 'Adverbs of Quantification', in E. Keenan (ed.), *Formal Semantics of Natural Language* (Cambridge, 1975), 3–15; repr. in his *Papers in Philosophical Logic* (Oxford, 1998), 5–20.

[119] Lynch, J., *Aristotle's School* (Berkeley, 1972).

[120] Makin, S., *Indifference Arguments* (Issues in Ancient Philosophy, 2; Oxford, 1993).

[121] Matthews, G., 'Senses and Kinds', *Journal of Philosophy*, 69 (1972), 149–57.

[122] Mueller, I., 'Aristotle on Geometrical Objects', *Archiv für Geschichte der Philosophie*, 52 (1970), 156–71; repr. in [92], 96–107.

[123] Owen, G. E. L., 'Logic and Metaphysics in some Earlier Works of Aristotle', in I. Düring and G. E. L. Owen, *Aristotle and Plato in the Mid-Fourth Century* (Göteborg, 1960), 163–90; repr. in [125], 180–99; also in [92], 13–32.

[124] —— 'Inherence', *Phronesis*, 10 (1965), 97–105; repr. in [125], 252–8.

[125] —— *Logic, Science and Dialectic: Collected Papers in Greek Philosophy*, ed. M. Nussbaum (London, 1986).

[126] Quine, W. Van O., *Word and Object* (Cambridge, Mass., 1960).

[127] Schofield, M., '*Metaphysics Z* 3: Some Suggestions', *Phronesis*, 17 (1972), 97–101.

[128] —— 'Explanatory Projects in *Physics*, 2. 3 and 7', *Oxford Studies in Ancient Philosophy*, suppl. vol. (1991), 29–40.

[129] Stroll, A., *Surfaces* (Minneapolis, 1988).

[130] Weyl, H., *Space–Time–Matter*, trans. H. L. Brose, 4th edn. (London, 1922).

[131] Wieland, W., 'The Problem of Teleology', in [91], 141–60.

[132] Wiggins, D., *Sameness and Substance* (Oxford, 1980).

VI. SOME MODERN DISCUSSIONS OF PLACE

[133] Cartwright, R., 'Scattered Objects', in K. Lehrer (ed.), *Analysis and Metaphysics* (Dordrecht, 1975), 153–71; repr. in his *Philosophical Papers* (Cambridge, 1987), 171–86. Introduces the notion of a receptacle in terms of topological concepts.

[134] CASATI, R., and VARZI, A., *Parts and Places: The Structures of Spatial Representation* (Cambridge, Mass., 1999).

[135] FORBES, G., 'Places as Possibilities of Location', *Noûs*, 21 (1987), 297–318. Elaborates a theory which supposedly gives formal teeth to Leibniz's comments on places.

[136] MOORE, G. E., *Commonplace Book 1919–1953*, ed. C. Lewy (London, 1962). See Notebook IV. 4; V. 22, 23, 25.

[137] O'SHAUGHNESSY, B., 'The Location of Sound', *Mind*, 66 (1957), 471–90.

[138] REICHENBACH, H., 'The Theory of Motion according to Newton, Leibniz, and Huyghens', in his *Modern Philosophy of Science: Selected Essays*, ed. and trans. M. Reichenbach (London, 1959), 46–66.

[139] RUSSELL, B., *ABC of Relativity* (London, 1925).

[140] SHARVY, R., 'Aristotle on Mixtures', *Journal of Philosophy*, 80 (1983), 439–57. Uses Cartwright [133] as a starting-point for mereo-topological musings on Aristotle.

[141] —— 'Mixtures', *Philosophy and Phenomenological Research*, 43 (1982–3), 227–39.

[142] STRAWSON, P. F., *Individuals* (London, 1959), esp. 218–24.

[143] TARSKI, A., 'Foundations of the Geometry of Solids', in his *Logic, Semantics, Metamathematics*, trans. J. Woodger (Oxford, 1956), 24–9. The foundational work of mereo-topology.

[144] TAYLOR, B., *Modes of Occurrence: Verbs, Adverbs and Events* (Aristotelian Society Series, 2; Oxford, 1985). See my Chapter 3, 99 n. 58.

[145] WIGGINS, D., 'The Individuation of Things and Places', *Proceedings of the Aristotelian Society*, suppl. vol. 37 (1963), 177–202.

[146] —— 'On Being in the Same Place at the Same Time', *Philosophical Review*, 77 (1968), 90–5.

[147] WOODS, M. J., 'The Individuation of Things and Places', *Proceedings of the Aristotelian Society*, suppl. vol. 37 (1963), 203–16.

Index Locorum

General Index